Recent Research in Polynomials

Edited by Faruk Özger

Published in London, United Kingdom

Recent Research in Polynomials
http://dx.doi.org/10.5772/intechopen.1001517
Edited by Faruk Özger

Contributors
Burcu Silindir, Ahmet Yantir, Sujit K. Bose, Imran Talib, Faruk Özger, Mustafa Abbas Fadhel, Gregory Natanson, Charles R. Johnson, Greyson C. Wesley, Xiaoyu Lin, Xiwen Liu, Sihan Zhou, Ohr Kadrawi, Vadim E. Levit, Ron Yosef, Matan Mizrachi, Adnan Ghazy AlAmosush, Mohd. Saif, Raghib Nadeem, Sarah Abdullah Qadha, Muneera Abdullah Qadha, Haibo Chen

First published in London, United Kingdom, 2023 by IntechOpen
IntechOpen is the global imprint of INTECHOPEN LIMITED, registered in England and Wales, registration number: 11086078, 5 Princes Gate Court, London, SW7 2QJ, United Kingdom

British Library Cataloguing-in-Publication Data
A catalogue record for this book is available from the British Library

Additional hard and PDF copies can be obtained from orders@intechopen.com

Recent Research in Polynomials
Edited by Faruk Özger
p. cm.
Print ISBN 978-1-83769-496-9
Online ISBN 978-1-83769-495-2
eBook (PDF) ISBN 978-1-83769-497-6

Meet the editor

Dr. Faruk Özger obtained his Ph.D. in Mathematics under the supervision of German mathematician Professor Eberhard Malkowsky in 2013. He has reviewed more than 100 articles and more than 50 national and international projects. He has published fifty research articles on applied and computational mathematics, operator theory, approximation theory, and functional analysis. Presently, he is a faculty member of the Department of Engineering Sciences, İzmir Katip Çelebi University, Türkiye.

Contents

Preface

The applications and importance of polynomials in the interdisciplinary field of mathematics and engineering are well known. Over the last several decades, experts have used polynomials to obtain pure and numerical results in many disciplines. They are incredibly useful mathematical tools because they are defined and can be calculated quickly on computer systems. While Umbral, Abel, Bell, Bernoulli, Euler, Boile, cyclotomic, Dickson, Fibonacci, and Touchard polynomials are used in algebra and combinatorics, Bernstein-type polynomials are widely used in approximation theory to expand its scope. Bernstein-type polynomials are also implemented to differential, integral, integro differential, differential, fractional integral, and fractional integrodifferential equations to obtain better numerical results and less error in approximation.

This book includes eleven chapters and devotes extensive coverage to recent developments in polynomials and their applications. In Chapter 1, Silindir and Yantir construct proper polynomials, namely delta and nabla generalized quantum polynomials, on (q, h)-time scales explicitly. In Chapter 2, Bose presents the application of algebraic methods for coding linear block codes and polynomials-based cyclic coding, focusing on the development of the Bose, Chaudhuri, Hocqenghem, and Reed–Solomon codes widely used in practice. In Chapter 3, Talib and Özger introduce the two new generalized operational matrices of Hermite polynomials, which are developed in the sense of the Riemann–Liouville fractional-order integral operator and Hilfer fractional-order derivative operator. In Chapter 4, Fadhel provides an overview of fitting parametric polynomials with control point coefficients. In Chapter 5, Natanson pays special attention to the rediscovery of Routh polynomials. In Chapter 6, Johnson et al. study the possible characteristic polynomials that may be realized by matrices A over a finite field such that the graph of A is G. In Chapter 7, Kadrawi et al. present a linear algorithm that uses dynamic programming to compute independence polynomials of trees. In Chapter 8, Alamosh introduces certain classes of bi-univalent functions by means of Gegenbauer polynomials and Hordam polynomials. In Chapter 9, Saif and Nadeem derive new explicit formulas and identities for poly-Changhee polynomials. Finally, in Chapter 10, Qadha et al. investigate the existence of limit cycles for quintic Kukles polynomial differential systems depending on a parameter.

Faruk Özger
Department of Engineering Sciences,
İzmir Katip Çelebi University,
İzmir, Türkiye

Chapter 1

Generalized Quantum Polynomials

Burcu Silindir and Ahmet Yantir

Abstract

On a general time scale, polynomials, Taylor's formula, and related subjects are described in terms of implicit and inefficient recursive relations. In this work, our primary goal is to construct proper polynomials, namely delta and nabla generalized quantum polynomials, on (q, h)-time scales explicitly. We show that generalized quantum polynomials play the same roles on (q, h)-time scales as ordinary polynomials play in $\mathbb{R}$ since they obey the additive properties and Leibnitz rules. Such polynomials which recover falling/rising and q-falling/q-rising factorials are constructed by the frame of forward and backward shifts. Additionally, we present delta- and nabla-Gauss' binomial formulas which provide many applications.

Keywords: (q, h)-time scales, delta generalized quantum polynomial, nabla generalized quantum polynomial, (q, h)-Taylor's formula, $(q$, h)-Gauss' binomial formula

1. Introduction

The polynomials occupy a very significant place in the theory of analysis. Once a polynomial is constructed, it is possible to express not only elementary functions but also some special functions in terms of infinite series of polynomials. On the other hand, a contribution to polynomials provides progress also on the theory of differential/difference equations because they could be proposed or solved by the frame of polynomials. Since the creation of modern calculus, the polynomials have been studied on continuous line or on discrete lattice where the stepsize h is constant, namely on $h\mathbb{Z}$

$$h\mathbb{Z} := \{hx : x \in \mathbb{Z}, \ \ h \in \mathbb{R}^+\}, \tag{1}$$

or on quantum numbers where the stepsize is not uniform, namely on $\mathbb{K}_q$

$$\mathbb{K}_q := \{q^n : \quad q \in \mathbb{R}, \quad q \neq 1, \quad n \in \mathbb{Z}\} \cup \{0\}. \tag{2}$$

After the invention of time scales by Stefan Hilger [1], many mathematical concepts constructed on discrete sets and continuous sets are unified without any big obstacles, especially in the theory of difference/differential equations. However, some key mathematical concepts such as polynomials and Taylor's formula have inefficient or implicit recursive definitions on general time scales and they are inapplicable in practice. Therefore, these concepts are studied not on a general time scale but instead on some specific time scales such as on (1) or on (2).

The so-called (q,h)-time scale $\mathbb{T}_{(q,h)}$ is introduced by Čermák and Nechvátal [2] as the unification of (1) and (2) in order to study fractional calculus. The importance of the (q,h)-time scale is beyond the expectations. In this special setting, it is possible to study the mathematical concepts which cannot be presented explicitly on a general time scale. In this work, we aim to present the proper forms of the polynomials on $\mathbb{T}_{(q,h)}$ in a manner that they assure the nature of (q,h)-time scales. In the literature, the studies on time scales proceed in two directions due to the delta and nabla derivative operators. For that reason, we present the unified polynomials on $\mathbb{T}_{(q,h)}$ in two separate forms: delta and nabla generalized quantum polynomials. We analyze the fundamental properties of both generalized quantum polynomials. One of the most significant advantage of studying on $\mathbb{T}_{(q,h)}$ is to attain the results which reduce to the results on $h\mathbb{Z}$, $\mathbb{K}q$ and $\mathbb{R}$ in the proper limits of h and q. We emphasize that the nabla generalized quantum polynomial unifies

1. nabla q-polynomial as $h \to 0$,
2. nabla h-polynomial as $q \to 1$,
3. ordinary polynomial as $(q,h) \to (1,0)$,

while the delta generalized quantum polynomial recovers

1. delta q-polynomial as $h \to 0$,
2. delta h-polynomial as $q \to 1$,
3. ordinary polynomial as $(q,h) \to (1,0)$.

The details of the above reductions under proper limits will be analyzed throughout the current work. Since the construction of $\mathbb{T}_{(q,h)}$ is well-defined, the reductions of forward and backward shifts, the nabla and delta derivatives, polynomials, Gauss Binomial formula to $h\mathbb{Z}$, $\mathbb{K}_q$ and $\mathbb{R}$ are consistent in both nabla and delta sense.

The current work is organized as follows. In Section 2, we give a very brief introduction to time scale calculus, define two appropriate (q,h)-time scales and we state the concepts of (q,h)-time scale calculus (in both nabla and delta sense). Section 3 is devoted to present the nabla generalized quantum polynomial equipped with its key features such as derivative rule and additive property. We also state and prove the nabla (q,h)-Taylor's formula from which we are able to establish the nabla (q,h)-Gauss binomial formula. In Section 4, we intensify on the delta generalized quantum polynomial. It is shown that the delta version of quantum binomial also possesses the key features of the polynomials. The delta version of (q,h)-Taylor's formula and Gauss binomial formula is presented. The notions and theories are explained by examples.

2. Preliminaries

A time scale $\mathbb{T}$ is a non-empty closed subset of $\mathbb{R}$. The most well-known examples of time scales: $\mathbb{Z}$, the Cantor set $\mathcal{C}$ and the discrete sets $h\mathbb{Z}$ (1) and $\mathbb{K}_q$ (2). We refer readers to see the pioneering article [1] and the books [3, 4] for the details of the theory of time scales and [5–7] for (q,h)-calculus. In this section, we only list the concepts of calculus on time scales, nabla calculus on $\mathbb{T}^1_{(q,h)}$ and delta calculus on $\mathbb{T}^2_{(q,h)}$ which we require throughout this work.

On an arbitrary time scale $\mathbb{T}$, the forward jump and the backward jump operators are given by

$$\sigma(x) := \inf\{s \in \mathbb{T} : s > x\}, \qquad \rho(x) := \sup\{s \in \mathbb{T} : s < x\}. \tag{3}$$

The Δ- and ∇-derivatives of a real-valued function on $\mathbb{T}$, are, respectively, defined by

$$f^{\Delta}(x) := \lim_{s \to x} \frac{f(\sigma(s)) - f(x)}{\sigma(s) - x}, \tag{4}$$

$$f^{\nabla}(x) := \lim_{s \to x} \frac{f(\rho(s)) - f(x)}{\rho(s) - x}. \tag{5}$$

For given $h, q \in \mathbb{R}^+$, we introduce two-parameter time scales by

$$\mathbb{T}^1_{(q,h)} := \{q^n x + [n]h : \quad n \in \mathbb{Z}\} \cup \left\{\frac{h}{1-q}\right\}, \quad q > 1, \quad x > \frac{h}{1-q}, \tag{6}$$

or

$$\mathbb{T}^2_{(q,h)} := \{q^n x + [n]h : \quad n \in \mathbb{Z}\} \cup \left\{\frac{h}{1-q}\right\}, \quad 0 < q < 1, \quad x < \frac{h}{1-q}, \tag{7}$$

where $[n]$ stands for the q-number given by

$$[n] := 1 + q + \cdots + q^{n-1}, \tag{8}$$

which tends to the non-negative integer n as $q \to 1$. It is clear that such time scales are generalizations of (1) and (2). On $\mathbb{T}^1_{(q,h)}$ and $\mathbb{T}^2_{(q,h)}$, the operators σ and ρ reduce to

$$\sigma(x) = qx + h, \qquad \rho(x) = \frac{x-h}{q}. \tag{9}$$

The point $\frac{h}{1-q}$ is an accumulation point, because of the limits

$$\lim_{n \to \infty} (q^n x + [n]h) = \frac{h}{1-q}, \quad 0 < q < 1 \tag{10}$$

and

$$\lim_{n \to \infty} q^{-n}(x - [n]h) = \frac{h}{1-q}, \quad q > 1. \tag{11}$$

The reason why we separately define (6) and (7) is to make our contributions consistent with the general theory of time scales.

Remark 2.1. It is also possible to define $\mathbb{T}^1_{(q,h)}$ for $0<q<1, \quad x<\frac{h}{1-q}$ and define $\mathbb{T}^2_{(q,h)}$ for $q>1, \quad x>\frac{h}{1-q}$. However, in those cases, the roles of the operators σ and ρ have to interchange in order to be consistent with the definitions (3).

Here we note that the time scales $\mathbb{T}^1_{(q,h)}$ and $\mathbb{T}^2_{(q,h)}$ are purely discrete sets with the exceptional accumulation point $\frac{h}{1-q}$. The nabla (q,h)-derivative and the delta (q,h)-derivative are defined as follows.

Definition 2.2. [7] Let $f(x):\mathbb{T}^1_{(q,h)}\rightarrow\mathbb{R}$ be a function. The *nabla (q,h)-derivative of* f is defined by

$$\tilde{D}_{(q,h)}f(x):=\begin{cases}\dfrac{f(x)-f\left(\dfrac{x-h}{q}\right)}{x-\left(\dfrac{x-h}{q}\right)} & \text{if} \quad x\neq\dfrac{h}{1-q},\\[2ex] \lim\limits_{s\rightarrow\left(\frac{h}{1-q}\right)^{+}}\dfrac{f(s)-f\left(\dfrac{h}{1-q}\right)}{s-\dfrac{h}{1-q}} & \text{if} \quad x=\dfrac{h}{1-q},\end{cases} \tag{12}$$

provided that the limits exist (see [3], Theorem 1.16 i).

Definition 2.3. [2, 5] Let $f(x):\mathbb{T}^2_{(q,h)}\rightarrow\mathbb{R}$ be a function. The *delta (q,h)-derivative* of f is defined by

$$D_{(q,h)}f(x):=\begin{cases}\dfrac{f(qx+h)-f(x)}{qx+h-x} & \text{if} \quad x\neq\dfrac{h}{1-q},\\[2ex] \lim\limits_{s\rightarrow\left(\frac{h}{1-q}\right)^{-}}\dfrac{f(s)-f\left(\dfrac{h}{1-q}\right)}{s-\dfrac{h}{1-q}} & \text{if} \quad x=\dfrac{h}{1-q},\end{cases} \tag{13}$$

provided that the limits exist (see [3], Theorem 1.16 i).

One of the most significant advantage of (q,h) time scales is to allow us to study q-calculus, h-calculus, and ordinary calculus on one hand. By the definition of jump operators (9), these operators reduce to

1. $\sigma(x)=\sigma_h(x)=x+h$ and $\rho(x)=\rho_h(x)=x-h$ as $q\rightarrow 1$.

2. $\sigma(x)=\sigma_q(x)=qx$ and $\rho(x)=\rho_q(x)=\frac{x}{q}$ as $h\rightarrow 0$.

3. $\sigma(x)=\rho(x)=x$ as $(q,h)\rightarrow(1,0)$.

The above reductions have very interesting consequences. If the above reductions are applied to the nabla (q,h)-derivative (12), we obtain the following list of reductions.

1. $\mathbb{T} = \mathbb{K}_q$: The nabla q-derivative $\tilde{D}_{(q,0)}f(x) = \nabla_q f(x)$.

2. $\mathbb{T} = h\mathbb{Z}$: The nabla h-derivative $\tilde{D}_{(1,h)}f(x) = \nabla_h f(x)$.

3. $\mathbb{T} = \mathbb{R}$: The ordinary derivative $\tilde{D}_{(1,0)}f(x) = \frac{df(x)}{dx}$.

Similarly, the delta (q,h)-derivative (13) has the below list of reductions.

1. $\mathbb{T} = \mathbb{K}_q$: The delta q-derivative [8] $D_{(q,0)}f(x) = \Delta_q f(x)$.

2. $\mathbb{T} = h\mathbb{Z}$: The nabla h-derivative $D_{(1,h)}f(x) = \Delta_h f(x)$.

3. $\mathbb{T} = \mathbb{R}$: The ordinary derivative $D_{(1,0)}f(x) = \frac{df(x)}{dx}$.

The above reductions are shown in **Figures 1** and **2**.

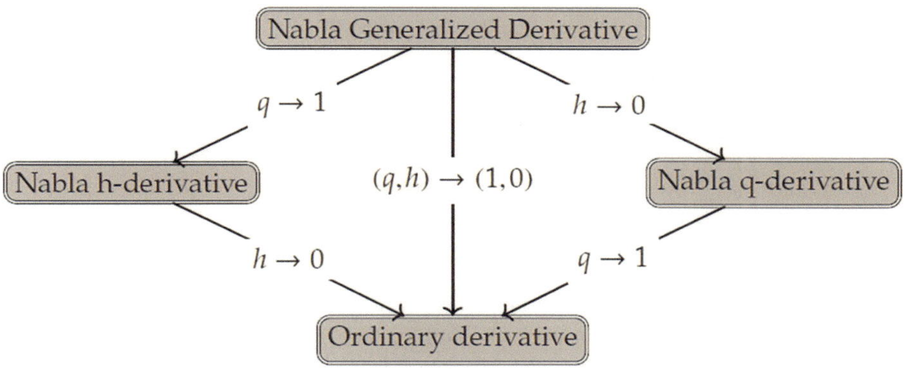

Figure 1.
Reductions of the nabla generalized derivative (or nabla (q,h)-derivative).

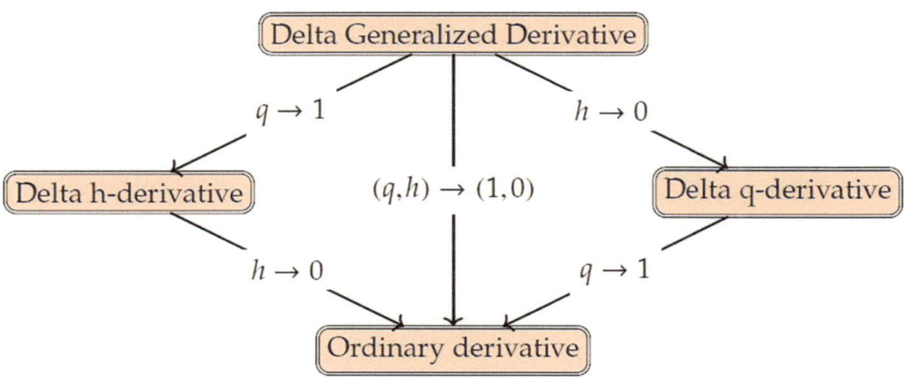

Figure 2.
Reductions of the delta generalized derivative (or delta (q,h)-derivative).

Proposition 2.4. [7] If $f,g:\mathbb{T}^1_{(q,h)}\to\mathbb{R}$ are any functions then the product rule for nabla (q,h)-derivative is given by

$$\tilde{D}_{(q,h)}(f(x)g(x)) = g(x)\tilde{D}_{(q,h)}f(x) + f\left(\frac{x-h}{q}\right)\tilde{D}_{(q,h)}g(x) \tag{14}$$

$$= f(x)\tilde{D}_{(q,h)}g(x) + g\left(\frac{x-h}{q}\right)\tilde{D}_{(q,h)}f(x). \tag{15}$$

Proposition 2.5. [5] If f,g are any real-valued functions defined on $\mathbb{T}^2_{(q,h)}$, then the product rule for delta (q,h)-derivative is expressed as

$$D_{(q,h)}(f(x)g(x)) = g(x)D_{(q,h)}f(x) + f(qx+h)D_{(q,h)}g(x) \tag{16}$$

$$= f(x)D_{(q,h)}g(x) + g(qx+h)D_{(q,h)}f(x). \tag{17}$$

Definition 2.6. We introduce the $\frac{1}{q}$-*number* as

$$[m]_{\frac{1}{q}} := 1 + \frac{1}{q} + \frac{1}{q^2} + \cdots + \frac{1}{q^{m-1}}, \quad m \in \mathbb{N}. \tag{18}$$

Similar to the q-number $[m]$, the $\frac{1}{q}$-*number* approaches to the non-negative number m as $q\to 1$. Moreover, there is a relation between q and $\frac{1}{q}$-numbers

$$[m]_{\frac{1}{q}} = \frac{[m]_q}{q^{m-1}}. \tag{19}$$

Definition 2.7. We introduce the $\frac{1}{q}$-*factorial* by

$$[m]_{\frac{1}{q}}! = [m]_{\frac{1}{q}}[m-1]_{\frac{1}{q}}\cdots[1]_{\frac{1}{q}}, \quad m \in \mathbb{N}, \tag{20}$$

while $[0]_{\frac{1}{q}}! := 1$.

Definition 2.8. We introduce the $\frac{1}{q}$-*binomial coefficient* as

$$\begin{bmatrix} m \\ j \end{bmatrix}_{\frac{1}{q}} := \frac{[m]_{\frac{1}{q}}!}{[m-j]_{\frac{1}{q}}![j]_{\frac{1}{q}}!}, \quad 0 \le j \le m, \quad m \in \mathbb{N}. \tag{21}$$

Proposition 2.9. The following properties hold for the $\frac{1}{q}$-binomial coefficient

$$\begin{bmatrix} m \\ i \end{bmatrix}_{\frac{1}{q}} = \begin{bmatrix} m \\ m-i \end{bmatrix}_{\frac{1}{q}}, \tag{22}$$

$$\begin{bmatrix} m \\ i \end{bmatrix}_{\frac{1}{q}} = q^{i(i-m)}\begin{bmatrix} m \\ i \end{bmatrix}_q. \tag{23}$$

Proof: By Definition 2.8, the property (i) is clear. For (ii), we employ the relation (19) which leads to

$$\begin{bmatrix} m \\ j \end{bmatrix}_{\frac{1}{q}} = \frac{[m]_{\frac{1}{q}}!}{[m-i]_{\frac{1}{q}}![i]_{\frac{1}{q}}!} = q^{i(i-m)} \frac{[m]_q!}{[m-j]_q![j]_q!} = q^{i(i-m)} \begin{bmatrix} m \\ i \end{bmatrix}_q. \tag{24}$$

3. Nabla generalized quantum polynomial

In this Section, our main aim is to present nabla (q,h)-analog of the polynomial $(x-\omega)^m$ on $\mathbb{T}^1_{(q,h)}$ (6) in a way that such a polynomial is consistent with the nabla (q,h)-derivative operator (12) and preserves the properties similar to ordinary polynomials. The findings of this section are based on the work [7].

Definition 3.1. We define the nabla generalized quantum polynomial (or nabla (q,h)-polynomial) by

$$(x-\omega)^{\tilde{m}}_{q,h} := \begin{cases} 1 & \text{if} \quad m=0, \\ \prod_{j=1}^{m}\left(x - \dfrac{\omega-[j-1]h}{q^{j-1}}\right) & \text{if} \quad m\in\mathbb{N}, \ \ \omega\in\mathbb{R}. \end{cases} \tag{25}$$

Example 3.2. We can demonstrate the nabla (q,h)-polynomial (25) for $m=4$

$$(x-\omega)^{\tilde{4}}_{q,h} = (x-\omega)\left(x - \frac{\omega-h}{q}\right)\left(x - \frac{\omega-[2]h}{q^2}\right)\left(x - \frac{\omega-[3]h}{q^3}\right). \tag{26}$$

i. If we set $h=\frac{1}{2}$ and $q=2$, then $[2]=1+q\to 3$ and $[3]=1+q+q^2\to 7$ and (26) turns out to be

$$(x-\omega)^{\tilde{4}}_{2,\frac{1}{2}} = (x-\omega)\left(x-\frac{\omega}{2}+\frac{1}{4}\right)\left(x-\frac{\omega}{4}+\frac{3}{8}\right)\left(x-\frac{\omega}{8}+\frac{7}{16}\right). \tag{27}$$

ii. If we set $h=\frac{1}{2}$ and $q=1$, then from (26) we have

$$(x-\omega)^{\tilde{4}}_{1,\frac{1}{2}} = (x-\omega)\left(x-\omega+\frac{1}{2}\right)(x-\omega+1)\left(x-\omega+\frac{3}{2}\right). \tag{28}$$

iii. If we set $h=0$ and $q=2$, then from (26) we derive

$$(x-\omega)^{\tilde{4}}_{2,0} = (x-\omega)\left(x-\frac{\omega}{2}\right)\left(x-\frac{\omega}{4}\right)\left(x-\frac{\omega}{8}\right). \tag{29}$$

Remark 3.3. We list the reductions of the nabla (q,h)-polynomial (25) as follows:

1. $\mathbb{T}=\mathbb{K}_q$: The nabla q-polynomial

$$(x-\omega)^{\tilde{m}}_{q,0} = (x-\omega)\left(x-q^{-1}\omega\right)\left(x-q^{-2}\omega\right)\cdots\left(x-q^{1-m}\omega\right). \tag{30}$$

2. $\mathbb{T}=h\mathbb{Z}$: The nabla h-polynomial

$$(x-\omega)^{\tilde{m}}_{1,h} = (x-\omega)(x-\omega+h)(x-\omega+2h)\cdots(x-\omega+(m-1)h). \tag{31}$$

3. $\mathbb{T} = \mathbb{R}$: The ordinary polynomial

$$(x-\omega)^{\tilde{m}}_{1,0} = (x-\omega)^m. \tag{32}$$

The reductions can be visualized in **Figure 3**.

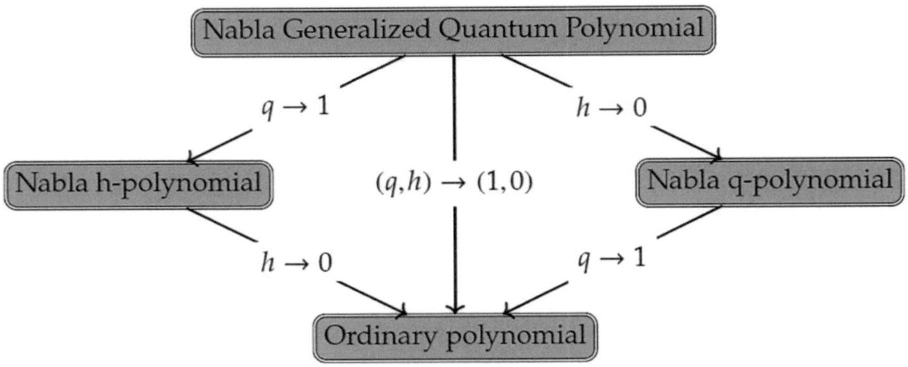

Figure 3.
Reductions of the nabla generalized quantum polynomial (or nabla (q,h)-polynomial).

Proposition 3.4. The Leibnitz rule for the nabla (q,h)-polynomial (25) is determined as

$$\tilde{D}^j_{(q,h)}(x-\omega)^{\tilde{m}}_{q,h} = [m]_{\frac{1}{q}}[m-1]_{\frac{1}{q}}\cdots[m-(j-1)]_{\frac{1}{q}}(x-\omega)^{\widetilde{m-j}}_{q,h}, \quad 1 \le j \le m. \tag{33}$$

Proof: For $j = 1$, we apply the nabla (q,h)-derivative on (25)

$$\tilde{D}_{(q,h)}(x-\omega)^{\tilde{m}}_{q,h} = \frac{(x-\omega)^{\tilde{m}}_{q,h} - \left(\frac{x-h}{q} - \omega\right)^{\tilde{m}}_{q,h}}{x - \dfrac{x-h}{q}}$$

$$= \left(\frac{x - \dfrac{\omega - [m-1]h}{q^{m-1}} - \dfrac{1}{q^{m-1}}\left(\dfrac{x-h}{q} - \omega\right)}{x - \dfrac{x-h}{q}}\right)(x-\omega)\left(x - \frac{\omega-h}{q}\right)\cdots\left(x - \frac{\omega - [m-2]h}{q^{m-2}}\right)$$

$$= \frac{1+q+\cdots q^{m-1}}{q^{m-1}}(x-\omega)^{\widetilde{m-1}}_{q,h} = \frac{[m]}{q^{m-1}}(x-\omega)^{\widetilde{m-1}}_{q,h}. \tag{34}$$

We obtain a very useful derivative rule for (25)

$$\tilde{D}_{(q,h)}(x-\omega)^{\tilde{m}}_{q,h} = [m]_{\frac{1}{q}}(x-\omega)^{\widetilde{m-1}}_{q,h} \tag{35}$$

For $j > 1$, we use (35) and immediately derive the property (33).

To be more precise, the nabla (q,h)-polynomial $(x-\omega)^{\tilde{m}}_{q,h}$ obeys a very significant derivative rule (35) as in ordinary calculus.

Theorem 1.1. Assume $\{P_0, P_1, \cdots P_M\}$ is a set of polynomials preserving the below conditions

1. $P_0(\omega) = 1$ and $P_k(\omega) = 0, k \in \mathbb{N}$,

2. $deg(P_k) = k, k \in \mathbb{N}_0$,

3. $D(P_k) = P_{k-1}, k \in \mathbb{N}$,

where D is any linear operator. Then Taylor's formula is presented by

$$Q(x) = \sum_{k=0}^{M} D^k Q(\omega) P_k(x), \tag{36}$$

where $Q(x)$ is any polynomial of degree M.

Proof: Consider the set of polynomials $A = \{P_0, P_1, \dots, P_M\}$ preserving the given conditions. Since $deg(P_k) = k$ for each k, then A is a linearly independent set of $M + 1$ polynomials. Assume V is a vector space of polynomials with dimension $M + 1$. Therefore A spans V and A turns out to be a basis for V. That is, any polynomial $Q(x)$ in V can be determined in terms of the elements of the basis A

$$Q(x) = \sum_{k=0}^{M} a_k P_k(x). \tag{37}$$

Using condition (i) on (37) leads to

$$Q(\omega) = \sum_{k=0}^{M} a_k P_k(\omega) = a_0 P_0(\omega) + a_1 P_1(\omega) + \cdots + a_M P_M(\omega) = a_0 P_0(\omega) = a_0. \tag{38}$$

We employ the linearity of D, the conditions (i) and (iii) to determine a_1

$$\begin{aligned} DQ(\omega) &= \sum_{k=0}^{M} a_k DP_k(\omega) \\ &= \sum_{k=1}^{M} a_k P_{k-1}(\omega) = a_1 P_0(\omega) + a_2 P_1(\omega) \cdots + a_M P_{M-1}(\omega) = a_1, \end{aligned} \tag{39}$$

and each coefficient a_m

$$\begin{aligned} D^m Q(\omega) &= \sum_{k=0}^{M} a_k D^m P_k(\omega) \\ &= \sum_{k=m}^{M} a_k P_{k-m}(\omega) = a_m P_0(\omega) + a_{m+1} P_1(\omega) + \cdots + a_M P_{M-m}(\omega) = a_m. \end{aligned} \tag{40}$$

As a conclusion, we end up with the desired Taylor's formula

$$Q(x) = \sum_{k=0}^{M} D^k Q(\omega) P_k(x). \tag{41}$$

Motivated by Theorem 1.1, we can state the following theorem.

Theorem 1.2. The nabla (q,h)-Taylor's formula is given by

$$Q(x) = \sum_{k=0}^{M} \tilde{D}^{k}_{(q,h)} Q(\omega) \frac{(x-\omega)^{k}_{\tilde{q,h}}}{[k]_{\frac{1}{q}}!}. \tag{42}$$

where $Q(x)$ is a polynomial of degree M.

Proof: Note that nabla (q,h)-derivative operator $\tilde{D}_{(q,h)}$ is linear and the set $\left\{1, (x-\omega)^{1}_{\tilde{q,h}}, \frac{(x-\omega)^{2}_{\tilde{q,h}}}{[2]_{\frac{1}{q}}!}, \dots, \frac{(x-\omega)^{M}_{\tilde{q,h}}}{[M]_{\frac{1}{q}}!}\right\}$ stands for a set of polynomials satisfying the properties of Theorem 1.1. Hence, the proof follows.

One of the most important distinguishing features of polynomials is the additive property. The nabla (q,h)-version of the additive property of nabla generalized quantum polynomials is stated as follows.

Proposition 3.5. The nabla (q,h)-polynomial (25) possesses the additive property.

$$(x-\omega)^{m+n}_{\tilde{q,h}} = (x-\omega)^{m}_{\tilde{q,h}} \cdot \left(x - \frac{\omega - [m]h}{q^m}\right)^{n}_{\tilde{q,h}}, \qquad m,n \in \mathbb{N}_0. \tag{43}$$

Proof: The proof is straightforward for $m = 0$ or $n = 0$ or both. For $m,n > 0$, the $(m+n)^{\text{th}}$ power of the delta (q,h)-polynomial (25) can be written as

$$\begin{aligned}(x-\omega)^{m+n}_{\tilde{q,h}} = (x-\omega)\left(x - \frac{\omega-h}{q}\right)\cdots\left(x - \frac{\omega-[m-1]h}{q^{m-1}}\right)\left(x - \frac{\omega-[m]h}{q^{m}}\right)\\ \cdots\left(x - \frac{\omega-[m+n-1]h}{q^{m+n-1}}\right).\end{aligned} \tag{44}$$

Note that the product of the first m terms is $(x-\omega)^{m}_{\tilde{q,h}}$. The product of the last n terms is $\left(x - \left(\frac{\omega-[m]h}{q^m}\right)\right)^{n}_{\tilde{q,h}}$ which is derived by replacing c by $\frac{\omega-[m]h}{q^m}$ in (25).

Example 3.6. Let us illustrate the additivity rule. Let $m = 3$, $n = 2$, then

$$\begin{aligned}(x-\omega)^{5}_{\tilde{q,h}} &= (x-\omega)^{3}_{\tilde{q,h}} \cdot \left(x - \frac{\omega-[3]h}{q^3}\right)^{2}_{\tilde{q,h}}\\ &= (x-\omega)\left(x - \frac{\omega-h}{q}\right)\left(x - \frac{\omega-[2]h}{q^2}\right)\left(x - \frac{\omega-[3]h}{q^3}\right)\left(x - \frac{\frac{\omega-[3]h}{q^3} - h}{q}\right)\\ &= (x-\omega)\left(x - \frac{\omega-h}{q}\right)\left(x - \frac{\omega-[2]h}{q^2}\right)\left(x - \frac{\omega-[3]h}{q^3}\right)\left(x - \frac{\omega-[4]h}{q^4}\right).\end{aligned} \tag{45}$$

i. If $h = 0$, then from (45) we have

$$\begin{aligned}(x-\omega)^{5}_{\tilde{q,0}} &= (x-\omega)^{3}_{\tilde{q,0}} \cdot \left(x - \frac{\omega}{q^3}\right)^{2}_{\tilde{q,0}}\\ &= (x-\omega)\left(x - q^{-1}\omega\right)\left(x - q^{-2}\omega\right)\left(x - q^{-3}\omega\right)\left(x - q^{-4}\omega\right).\end{aligned} \tag{46}$$

ii. If $q = 1$, then (45) implies

$$\begin{aligned}(x-\omega)^{5}_{\widetilde{1,h}} &= (x-\omega)^{3}_{\widetilde{1,h}} \cdot (x-\omega+3h)^{2}_{\widetilde{1,h}} \\ &= (x-\omega)(x-\omega+h)(x-\omega+2h)(x-\omega+3h)(x-\omega+4h).\end{aligned} \tag{47}$$

The nabla (q,h)-version of the celebrating Gauss' Binomial formula is as follows:

Theorem 1.3. On $\mathbb{T}^{1}_{(q,h)}$, the nabla (q,h)-analog of Gauss' Binomial formula can be presented in two equivalent forms

$$(x-\omega)^{m}_{\widetilde{q,h}} = \sum_{j=0}^{m} \begin{bmatrix} m \\ j \end{bmatrix}_{\frac{1}{q}} (0-\omega)^{m-j}_{\widetilde{q,h}} \cdot (x-0)^{j}_{\widetilde{q,h}} = \sum_{j=0}^{m} \begin{bmatrix} m \\ j \end{bmatrix}_{\frac{1}{q}} (0-\omega)^{j}_{\widetilde{q,h}} \cdot (x-0)^{m-j}_{\widetilde{q,h}}. \tag{48}$$

Proof: If we set $f(x) := (x-\omega)^{m}_{\widetilde{q,h}}$, by the use of Proposition 3.4, we can calculate

$$\tilde{D}^{j}_{(q,h)} f(0) = [m]_{\frac{1}{q}} [m-1]_{\frac{1}{q}} \cdots [m-(j-1)]_{\frac{1}{q}} (0-\omega)^{m-j}_{\widetilde{q,h}}, \quad 0 \leq j \leq m. \tag{49}$$

As a consequence of Theorem 1.2, we conclude that

$$(x-\omega)^{m}_{\widetilde{q,h}} = \sum_{j=0}^{m} \frac{\tilde{D}^{j}_{(q,h)} f(0)(x-0)^{j}_{\widetilde{q,h}}}{[j]_{\frac{1}{q}}!} = \sum_{j=0}^{m} \begin{bmatrix} m \\ j \end{bmatrix}_{\frac{1}{q}} (0-\omega)^{m-j}_{\widetilde{q,h}} (x-0)^{j}_{\widetilde{q,h}}. \tag{50}$$

For the second equivalent form, we use Proposition 2.9 (i) to rewrite

$$(x-\omega)^{m}_{\widetilde{q,h}} = \sum_{j=0}^{m} \begin{bmatrix} m \\ m-j \end{bmatrix}_{\frac{1}{q}} (0-\omega)^{m-j}_{\widetilde{q,h}} \; (x-0)^{j}_{\widetilde{q,h}} = \sum_{k=0}^{m} \begin{bmatrix} m \\ k \end{bmatrix}_{\frac{1}{q}} (0-\omega)^{k}_{\widetilde{q,h}} \; (x-0)^{m-k}_{\widetilde{q,h}}, \tag{51}$$

where we used the index change $k = m - j$.

Example 3.7. Consider $m = 3$. Then one can compute

$$\begin{aligned}(x-\omega)^{3}_{\widetilde{q,h}} &= \sum_{j=0}^{3} \begin{bmatrix} 3 \\ j \end{bmatrix}_{\frac{1}{q}} (0-\omega)^{3-j}_{\widetilde{q,h}} \cdot (x-0)^{j}_{\widetilde{q,h}} \\ &= (0-\omega)^{3}_{\widetilde{q,h}} + [3]_{\frac{1}{q}}(0-\omega)^{2}_{\widetilde{q,h}} \cdot (x-0)^{1}_{\widetilde{q,h}} + [3]_{\frac{1}{q}}(0-\omega)^{1}_{\widetilde{q,h}} \cdot (x-0)^{2}_{\widetilde{q,h}} + (x-0)^{3}_{\widetilde{q,h}} \\ &= (x-\omega)\left(x - \frac{\omega-h}{q}\right)\left(x - \frac{\omega-[2]h}{q^2}\right).\end{aligned} \tag{52}$$

4. Delta generalized quantum polynomial

In this Section, we present delta (q,h)-analog of the polynomial $(x-\omega)^m$ on $\mathbb{T}^{2}_{(q,h)}$ (7). Such polynomial satisfies the derivative rule with respect to the delta (q,h)-derivative operator (13) and additive property. The results of this section are based on the articles [5] and [6].

Definition 4.1. We define the delta generalized quantum polynomial (or delta (q,h)-polynomial) by

$$(x-\omega)_{q,h}^{m} := \begin{cases} 1 & \text{if} \quad n=0, \\ \prod_{j=1}^{m}\left(x-q^{j-1}\omega-[j-1]h\right) & \text{if} \quad m\in\mathbb{N}, \ \ \omega\in\mathbb{R}. \end{cases} \tag{53}$$

Example 4.2. We can demonstrate the delta (q,h)-polynomial (53) for $m=3$

$$(x-\omega)_{q,h}^{3} = (x-\omega)(x-q\omega-h)\left(x-q^{2}\omega-[2]h\right). \tag{54}$$

i. If we set $h=\frac{1}{3}$ and $q=\frac{1}{2}$, then $[2]=1+q\rightarrow\frac{3}{2}$ from which (54) becomes

$$(x-\omega)_{\left(\frac{1}{2},\frac{1}{3}\right)}^{3} = (x-\omega)\left(x-\frac{\omega}{2}-\frac{1}{3}\right)\left(x-\frac{\omega}{4}-\frac{1}{2}\right). \tag{55}$$

ii. If we set $h=\frac{1}{2}$ and $q=1$, then (54) yields as

$$(x-\omega)_{\left(1,\frac{1}{3}\right)}^{3} = (x-\omega)\left(x-\omega-\frac{1}{2}\right)(x-\omega-1). \tag{56}$$

iii. If we set $h=0$ and $q=\frac{1}{2}$, then (54) leads to

$$(x-\omega)_{\left(\frac{1}{2},0\right)}^{3} = (x-\omega)\left(x-\frac{\omega}{2}\right)\left(x-\frac{\omega}{4}\right). \tag{57}$$

Remark 4.3. We list the reductions of the delta (q,h)-polynomial (53) as follows:

1. $\mathbb{T}=\mathbb{K}_q$: The delta q-polynomial [8]

$$(x-\omega)_{q,0}^{m} = (x-\omega)(x-q\omega)\left(x-q^{2}\omega\right)\cdots\left(x-q^{m-1}\omega\right). \tag{58}$$

2. $\mathbb{T}=h\mathbb{Z}$: The delta h-polynomial

$$(x-\omega)_{1,h}^{m} = (x-\omega)(x-\omega-h)(x-\omega-2h)\cdots(x-\omega-(m-1)h). \tag{59}$$

3. $\mathbb{T}=\mathbb{R}$: The ordinary polynomial (**Figure 4**)

$$(x-\omega)_{1,0}^{m} = (x-\omega)^{m}. \tag{60}$$

Proposition 4.4. The delta (q,h)-polynomial (53) obeys the Leibniz formula

$$D_{(q,h)}^{j}(x-\omega)_{q,h}^{m} = [m][m-1]\cdots[m-(j-1)](x-\omega)_{q,h}^{m-j}, \quad 1\leq j\leq m. \tag{61}$$

Proof: For $j=1$, the property (61) holds since

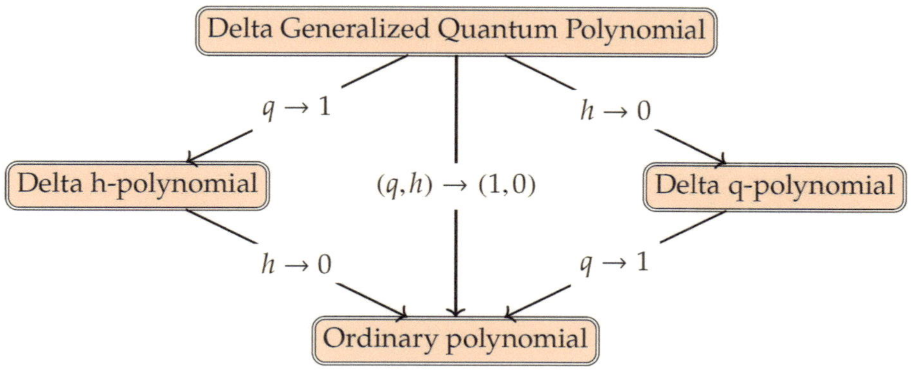

Figure 4.
Reductions of the delta generalized quantum polynomial (or delta (q,h)-polynomial).

$$\begin{aligned} D_{(q,h)}(x-\omega)^m_{q,h} &= \frac{(qx+h-\omega)^m_{q,h} - (x-\omega)^m_{q,h}}{qx+h-x} \\ &= (x-\omega)^{m-1}_{q,h}\left(\frac{q^{m-1}(qx+h-\omega) - (x-(q^{m-1}\omega+[m-1]h))}{(q-1)x+h}\right) \\ &= (x-\omega)^{m-1}_{q,h}\left(\frac{(q^m-1)x+q^{m-1}h+[m-1]h}{(q-1)x+h}\right) \\ &= (x-\omega)^{m-1}_{q,h}\left(\frac{[m](q-1)x+[m]h}{(q-1)x+h}\right) = [m](x-\omega)^{m-1}_{q,h}. \end{aligned} \tag{62}$$

The formula (61) yields by applying the delta (q,h)-derivative to the delta (q,h)-polynomial (53) j-times successively.

Proposition 4.5. The delta delta (q,h)-polynomial (53) admits the additive identity

$$(x-\omega)^{m+n}_{q,h} = (x-\omega)^m_{q,h}\cdot(x-(q^m\omega+[m]h))^n_{q,h}, \qquad m,n\in\mathbb{N}_0 \tag{63}$$

Proof: The identity is trivial if $m=0$ or $n=0$ or both. If $m,n>0$ the definition of the delta (q,h)-polynomial (53) allows us to write

$$\begin{aligned} (x-\omega)^{m+n}_{q,h} &= (x-\omega)(x-q\omega-h)\cdots(x-q^{m-1}\omega-[m-1]h)(x-q^m\omega-[m]h) \\ &= (x-\omega)^m_{q,h}\cdot\underbrace{(x-q^m\omega-[m]h)(x-q^{m+1}\omega-[m+1]h)\cdots(x-q^{m+n-1}\omega-[m+n-1]h)} \end{aligned} \tag{64}$$

where the underbraced term is nothing but $(x-(q^m\omega+[m]h))^n_{q,h}$ which is obtained by replacing ω by $q^m\omega+[m]h$ in (53).

Example 4.6. Here we illustrate the additivity property. Let $m=2$, $n=2$, then

$$\begin{aligned} (x-\omega)^4_{q,h} &= (x-\omega)^2_{q,h}\cdot(x-q^2\omega-[2]h)^2_{q,h} \\ &= (x-\omega)(x-q\omega-h)(x-q^2\omega-[2]h)(x-(q(q^2\omega+[2]h)+h)) \\ &= (x-\omega)(x-q\omega-h)(x-q^2\omega-[2]h)(x-q^3\omega-[3]h). \end{aligned} \tag{65}$$

i. If $h = 0$, then from (65) we have

$$(x-\omega)^{4}_{q,0} = (x-\omega)^{2}_{q,0} \cdot \left(x-q^{2}\omega\right)^{2}_{q,0} = (x-\omega)(x-q\omega)\left(x-q^{2}\omega\right)\left(x-q^{3}\omega\right). \quad (66)$$

ii. If $q = 1$, then (65) implies

$$(x-\omega)^{4}_{1,h} = (x-\omega)^{2}_{1,h} \cdot (x-\omega-2h)^{2}_{1,h} = (x-\omega)(x-\omega-h)(x-\omega-2h)(x-\omega-3h). \quad (67)$$

Theorem 1.4. The delta (q,h)-analogue of Taylor's formula is given by

$$Q(x) = \sum_{k=0}^{M} D^{k}_{(q,h)} Q(\omega) \frac{(x-\omega)^{k}_{q,h}}{[k]!}. \quad (68)$$

where $Q(x)$ is a polynomial of degree M.

Proof: The proof is based on Theorem 1.1. Since delta (q,h)-derivative operator is linear and the set $\left\{1, (x-\omega)^{1}_{q,h}, \frac{(x-\omega)^{2}_{q,h}}{[2]!}, \dots, \frac{(x-\omega)^{M}_{q,h}}{[M]!}\right\}$ stands for a set of polynomials satisfying the properties of Theorem 1.1. Therefore, the proof finishes.

Theorem 1.5. On $\mathbb{T}^{2}_{(q,h)}$, the delta (q,h)-analog of the Gauss Binomial formula has the following equivalent forms

$$(x-\omega)^{m}_{q,h} = \sum_{k=0}^{m} \begin{bmatrix} m \\ k \end{bmatrix} (0-\omega)^{m-k}_{q,h} \cdot (x-0)^{k}_{q,h} = \sum_{k=0}^{m} \begin{bmatrix} m \\ k \end{bmatrix} (0-\omega)^{k}_{q,h} \cdot (x-0)^{m-k}_{q,h}. \quad (69)$$

Proof: We choose $f(x) = (x-\omega)^{m}_{q,h}$ and employ Theorem 1.4 about $\omega = 0$ to obtain

$$(x-\omega)^{m}_{q,h} = \sum_{k=0}^{m} D^{k}_{(q,h)} f(0) \frac{(x-0)^{k}_{q,h}}{[k]!}. \quad (70)$$

By Proposition 4.4, we derive

$$D^{k}_{(q,h)} f(0) = [m][m-1]\cdots[m-(k-1)](0-\omega)^{m-k}_{q,h}, \quad 0 \le k \le m. \quad (71)$$

Therefore, (70) leads to the delta (q,h)-Gauss Binomial formula

$$(x-\omega)^{m}_{q,h} = \sum_{k=0}^{m} \begin{bmatrix} m \\ k \end{bmatrix} (0-\omega)^{m-k}_{q,h} \cdot (x-0)^{k}_{q,h}. \quad (72)$$

Since $\begin{bmatrix} m \\ k \end{bmatrix} = \begin{bmatrix} m \\ m-k \end{bmatrix}$, one can use the index change $i = m-k$ and end up with the second form

$$(x-\omega)^{m}_{q,h} = \sum_{k=0}^{m} \begin{bmatrix} m \\ m-k \end{bmatrix} (0-\omega)^{m-k}_{q,h} \cdot (x-0)^{k}_{q,h} = \sum_{i=0}^{m} \begin{bmatrix} m \\ i \end{bmatrix} (0-\omega)^{i}_{q,h} \cdot (x-0)^{m-i}_{q,h}. \quad (73)$$

Example 4.7. Let $m = 3$. We may calculate

$$\begin{aligned}(x-\omega)^3_{q,h} &= \sum_{k=0}^{3}\begin{bmatrix}3\\k\end{bmatrix}(0-\omega)^{3-k}_{q,h}\cdot(x-0)^k_{q,h}\\ &= (0-\omega)^3_{q,h} + [3](0-\omega)^2_{q,h}(x-0)^1_{q,h} + [3](0-\omega)^1_{q,h}(x-0)^2_{q,h} + (x-0)^3_{q,h}\\ &= (x-w)(x-qw-h)\left(x-q^2w-[2]h\right).\end{aligned} \tag{74}$$

5. Conclusions

We presented delta and nabla generalized quantum polynomials which are determined by the use of forward and backward shifts. We showed that such polynomials recover delta q-, delta h-, nabla q-, nabla h- and ordinary polynomials. We emphasize that the study on generalized quantum polynomials not only unify polynomials (and related subjects) on h-lattice sets, quantum numbers and on $\mathbb{R}$ but also create a paradigm on the theory of special functions (power functions [9], hypergeometric functions, Bernstein polynomials, Bernoulli polynomials, etc.) and combinatorics.

Author details

Burcu Silindir[1*†] and Ahmet Yantir[2†]

1 Faculty of Science, Department of Mathematics, Dokuz Eylül University, İzmir, Türkiye

2 Faculty of Science and Letters, Department of Mathematics, Yaşar University, İİzmir, Türkiye

*Address all correspondence to: burcu.silindir@deu.edu.tr

† These authors contributed equally.

References

[1] Hilger S. Analysis on measure chains–A unified approach to continuous and discrete calculus. Results in Mathematics. 1990;**18**:18-56

[2] Čermák J, Nechvátal L. On (q,h)-analogue of fractional calculus. Journal of Nonlinear Mathematical Physics. 2010;**17**(1):51-68

[3] Bohner M, Peterson A. Dynamic Equations on Time Scales. Boston, USA: Birkhauser; 2001

[4] Bohner M, Peterson A. Advances in Dynamic Equations on Time Scales. Boston, USA: Birkhauser; 2003

[5] Silindir B, Yantir A. Generalized quantum exponential function and its applications. Univerzitet u Nišu. 2019; **33**(15):907-4922

[6] Silindir B, Yantir A. Gauss's binomial formula and additive property of exponential functions on $\mathbb{T}_{(q,h)}$. Univerzitet u Nišu. 2021;**35**(11): 3855-3877

[7] Yantir A, Silindir B, Tuncer Z. Bessel equation and Bessel function on $\mathbb{T}_{(q,h)}$. Turkish Journal of Mathematics. 2022; **(46)8**:3300-3322

[8] Kac V, Cheung P. Quantum Calculus. New York: Springer; 2002

[9] Gergün S, Silindir B, Yantir A. Power function and Binomial series on $\mathbb{T}_{(q,h)}$. Advanced Mathematics in Science and Engineering. 2022

Chapter 2

Application of Polynomials in Coding of Digital Data

Sujit K. Bose

Abstract

Communication of information is nowadays ubiquitous in the form of words, pictures and sound that require digital conversion into binary coded signals for transmission due to the requirement of electronic circuitry of the implementing devices and machines. In a subtle way, polynomials play a very deep, important role in generating such codes without errors and distortion over long multiple pathways. This chapter surveys the very basics of such polynomials-based coding for easy grasp in the realization of such a complicated technologically important task. The coding is done in terms of just 0 and 1 that entails use of the algebra of finite Galois fields, using polynomial rings in particular. Even though more than six decades have passed since the foundations of the subject of the theory of coding were laid, the interest in the subject has not diminished as is apparent from the continued appearance of books on the subject. Beginning with the introduction of the ASCII codification of alphanumeric and other symbols developed for early generation computers, this article proceeds with the application of algebraic methods of coding of linear block codes and the polynomials-based cyclic coding, leading to the development of the BCH and the Reed–Solomon codes, widely used in practices.

Keywords: coding, digital data, finite field, polynomial, cyclic codes

1. Introduction

Communication from a "source" to a "receiver" forms a vast buzz of activity over the entire globe. Physically, it is carried out as digital signals transmitted via different pathways such as light pulses through fiber-optic cables and radiowaves through air and even over the outer space. The digital signals are created in *bits* of pulses by electronic circuitry that work on the principle of (nearly) 0 and + 5 volts (Leach et al. [1], p. 3). A handy mathematical representation of the two voltages is just the set of two elements {0, 1}. An information thus consists of a string of 0's and 1's carried optically or electromagnetically through air or outer space as the case may be. The physical channel of information transmission is naturally noisy that was treated mathematically by Claude Shannon in a seminal paper [2] entitled "A mathematical theory of communication", showing that in such noisy channel, there is a number called *channel capacity* such that reliable communication is achieved at any rate below the channel capacity by proper encoding and decoding techniques of any information.

This paper marked the beginning of the subject of Coding Theory for encoding and decoding of information through the maze of channels.

In its widespread usage, information can be literal or numeric. It can also be audio or video, all encoded in $\{0, 1\}$ bits. Moreover, special encoding and decoding is required for compression of the data at the very source to reduce the volume, storage on hard disks and encryption and decryption to maintain security of the data to be transmitted. The subject is therefore vast, and the mathematics involved is very special over the two numbers 0 an 1, studied in Abstract Algebra as a very special example of a *finite field*. However, according to Richard Hamming, a pioneer of the subject, "Mathematics is an interesting intellectual sport but it should not be allowed to stand in the way of obtaining sensible information about physical processes". In that spirit, this article is aimed to provide just the flavor of introducing the use of polynomials over the $\{0, 1\}$ field for developing a few practical methods of coding. No attempt is made to describe the corresponding methods of decoding, keeping in view the scope of this article. Detailed account of the subject and other methods of coding not treated here can be found in the texts by Bose [3], Kythe and Kythe [4], Roth [5], Moon [6], Ling and Xing [7], Blahut [8], Gathen and Gerhard [9], Proakis and Salehi [10], Adámek [11], and Lin and Costello [12]. All of these texts deal with polynomials over finite Galois fields to varying degrees of detail for the development of important codes like the practically important cyclic codes introduced by Prange [13], the BCH codes discovered by Bose, Ray-Chaudhuri [14] and independently by Hocqenghem [15], and the RS codes developed by Reed and Solomon [16]. The texts cited above also describe in detail, decoding methods of coded messages by a receiver, using special polynomials called *syndromes*. Besides the polynomials-based coding methods, the texts also give accounts of further development of codes with memory of past code words (convolutional coding) and codes for modulation of data transmission. Information theoretic probabilistic uncertainties of channelizing data transmission are also discussed in these texts to considerable extent. The list of texts is indicative of the continued interest in the topic of digital coding even now, ever since the appearance of the paper by Shannon. In what follows, Sections 2, 3, and 4 may be considered as preparatory coding methods before the appearance of polynomials over finite fields.

2. Coding

Transmission of a *message* of an information from a source to a destination is broadly classified into two categories: *source coding* of the message and *channel coding* for transmission through a channel. Both the types of coding are done mostly in terms of *bits* 0 and 1, because of the requirement of electronic implementation. Errors may occur at the source itself as well as in the transmission channel, and both of them require rectification in the transmitted message. The coding must be such that it can be uniquely decoded.

As an example of source coding, suppose that there is a message consisting of decimal numbers 0, 1, 2, 3, 4, ... and alphabets A, B, C, D, ... The first list can be bit converted by striking off the numbers 2, 3, ..., and 9. So that the binary code of 0, 1, 2, 3, 4, ... becomes 0, 1, 10, 11, 100, ... Coding a moderately large decimal number would be prohibitively very large and may not be uniquely decodable. For instance, the message 1011 could be decoded as 23 or 51 or 211. Giving a *place value* to the bits in the list as in the case of decimal numbers, a binary number listed as $a_0a_1a_2a_3\cdots$ can, however, be uniquely converted to its decimal equivalent by the *polynomial expression*

$$(a_0a_1a_2a_3\cdots)_2 = \left(\cdots\cdots a_3 \times 2^3 + a_2 \times 2^2 + a_1 \times 2^1 + a_0 \times 2^0\right)_{10} \tag{1}$$

where the suffixes 2 and 10 on the two sides of Eq. (1) indicate the concerned number to be binary or decimal as bases. Thus, the decimal equivalent of the binary

$$1011 = 1 \times 2^3 + 0 \times 2^2 + 1 \times 2^1 + 1 \times 2^0 = 8 + 0 + 2 + 1 = 11 \tag{2}$$

But such a method is not applicable in the case of the alphabets A, B, C, D,......

Electronic hardware considerations on the other hand are realized for unique decoding using *linear block codes*, in which every numeral and alphabet is written in fixed length block of bits called *word*. In this respect the ASCII code of *seven bits* is important. For succinct coding, *octal codes* are used, in which the base is 8 consisting of the numbers 0, 1, 2, 3, 4, 5, 6, and 7 striking off the decimals 8 and 9. It can be easily verified that the octal digits can be represented by blocks of 3 bits as the two bits can be permuted in $2^3 = 8$

Octal digits	**0**	**1**	**2**	**3**	**4**	**5**	**6**	**7**
Binary blocks	000	001	010	011	100	101	110	111

The seven places of an ASCII word can be filled in $2^7 = 128$ ways by the two bits 0 and 1. Accordingly 128 symbols, alphanumeric or any other can be bit coded. A very short list of ASCII words is presented below:

0	**1**	**2**	**3**	**A**	**B**	**C**	**<**	**=**	**>**
$0(60)_8$	$0(61)_8$	$0(62)_8$	$0(63)_8$	$1(01)_8$	$1(02)_8$	$1(03)_8$	$0(74)_8$	$0(75)_8$	$0(76)_8$

A full table of the code is given in Adámek [8], p.10. In practice one additional bit place is kept with every ASCII word as a check, so that the word length is actually 8. An 8 bit word is called a byte. A computer usually uses a word length of 4 bytes or 32 bits. For representation in such long length, a base of 16 = 2^4 bits is used to represent each symbol by 4 bits. Such coding is called hexadecimal represented by the 16 symbols 0, 1, 2,....., 9, A, B, C, D, E, F.

Audio and video information are continuous analog information. Such signals are *sampled* (see Leach et al. [1], p. 3) at small time or space–time intervals and are thus rendered discrete to generate digital data and suitably coded in bits. Foe instance, the basic colors of red, green, and blue are coded as #FF0000, #00FF00, and #0000FF, respectively, where the code of the symbol # is $0(43)_8$. The respective color codes of white and black are #FFFFFF and #000000.

3. Algebraic formulation

A digital *Code C* is a sequence of *words* constituted of string of bits of some fixed length n. A code word C can therefore be considered as an $n-$vector denoted as $\mathbf{a} = a_0a_1a_2\cdots a_{n-1}$ in which the elements $a_0, a_1, a_2, \cdots, a_{n-1} \in (0,1)$. The commas separating the elements of $\mathbf{a}$ are dropped as unwanted symbol to form the block of bits. A code word of C usually carries the *message* bits and some extra bits for detection of errors and their correction. This is necessary even though the loading time of bits increases to some extent. If the message consists of k bits, and $n-k$ bits for error detection and correction, then it is called an (n, k) code. Evidently by permutation, the number of messages in C that can be formed is 2^k. As an example the $(7, 4)$ code of three words

1000011   0100101   0010110

with the last three bits 011, 101, 110 of each message word represents the decimal number 842 according to the 4 bit hexadecimal representation of the three decimal digits.

A code C when transmitted through a channel may contain some error and transmitted as $b_0b_1b_2 \cdot b_{n-1}$ instead of the actual code $a_0a_1a_2\cdots a_{n-1}$. For error detection and correction, the following definition named after Richard Hamming [17] is introduced to keep the code words to be as wide apart as possible.

Definition (Hamming distance). Given two words $\mathbf{a} = a_0a_1a_2\cdots a_{n-1}$and $\mathbf{b} = b_0b_1b_2\cdots b_{n-1}$ their distance $d(\mathbf{a}, \mathbf{b})$ is defined as the number of positions in which **a** and **b** differ. Thus,

$$d(\mathbf{a}, \mathbf{b}) = \text{number of indices if or which}\, a_i \neq b_i, (i = 0, 1, 2, \cdots, n-1) \qquad (3)$$

As an example consider three words $\mathbf{x} = 1000011$, $\mathbf{y} = 0100101$, and $\mathbf{z} = 0010110$ of the preceding example, then $d(\mathbf{x}.\mathbf{y}) = d(\mathbf{y}, \mathbf{z}) = d(\mathbf{z}, \mathbf{x}) = 4$. But **x** and **y** as before and $\mathbf{z} = 1100110$, one gets $d(\mathbf{x}, \mathbf{y} = 4$, $d(\mathbf{y}, \mathbf{z}) = d(\mathbf{z}, \mathbf{x}) = 3$. The 4 bit message of the example in decimals is $x = 8, y = 4, z = 2$ while $z = C$ (in Hex) or decimal 12, in the second case. It may be observed that $d(\mathbf{x}, \mathbf{y}) \leq d(\mathbf{y}, \mathbf{z}) + d(\mathbf{z}, \mathbf{x})$ and $d(\mathbf{z}, \mathbf{x}) \leq d(\mathbf{x}, \mathbf{y}) + d(\mathbf{y}, \mathbf{z})$. In general, it can easily be shown that the Hamming distance $d(\mathbf{a}, \mathbf{b})$ is a metric on the set of words of length n satisfying the triangle inequality (Adámek [11], p. 46), as demonstrated in the example.

The definition of Hamming distance is useful for detection of errors in the following manner. A block code C is said to *detect t errors* provided that for each code word **a** and each word **b** obtained from **a** by corrupting $1, 2, \cdots, t$ symbols **b** is not a code word.

Definition. The minimum distance $d(C)$ of a code C is the smallest Hamming distance of two distinct code words of the code C, that is,

$$d(C) = \min\{d(\mathbf{a}, \mathbf{b}) \mid \mathbf{a}, \mathbf{b}\, \text{are word sin}\, C \text{and}\, \mathbf{a} \neq \mathbf{b}\} \qquad (4)$$

For example suppose that $C = [1000011\, 0100101\, 0010110], d(C) = 4$ and for $= [1000011\, 0100101\, 1100110], d(C) = 3$. The following proposition deals with the question of detection of errors in a code.

Proposition 1. A code C detects t errors if and only if $d(C) > t$.

Proof. If $d(C) \leq t$, then C does not detect t errors. In fact, let **a**, **a'** be correct and received code words with $d(\mathbf{a}, \mathbf{a}') = d(C)$. Then $d(\mathbf{a}, \mathbf{a}') \leq t$, so the error which changes the original code **a** to the received word **a'** escapes undetected. On the other hand, if $d(C) > t$, then C detects t errors. For, by definition, $1 \leq d(\mathbf{a}, \mathbf{a}') \leq t$. Then **a'** can not be a code word since $d(C) \leq d(\mathbf{a}, \mathbf{a}') \leq t$.

Regarding correction of codes, let **a'** be the word obtained by corrupting $1, 2, \cdots, t$ bits of the code word **a**, then the Hamming distance $d(\mathbf{a}, \mathbf{a}')$ is strictly smaller than that between **a'** and any other code word **b**, that is, $d(\mathbf{a}, \mathbf{a}') < d(\mathbf{b}, \mathbf{a}')$. This leads to the following:

Proposition 2. A code C corrects t errors if and only if $d(C) \geq 2t + 1$.

Proof. Let $d(C) \geq 2t + 1 > 2t$, where $d(\mathbf{a}, \mathbf{a}') \leq t$. Hence for any other code word $\mathbf{b} \neq \mathbf{a} : d(\mathbf{a}, \mathbf{b}) \geq d(C) > 2t$, and by the triangle inequality

$$d(\mathbf{a}, \mathbf{a}') + d(\mathbf{a}', \mathbf{b}) \geq d(\mathbf{a}, \mathbf{b}) > 2t$$

Hence,

$$d(\mathbf{a}', \mathbf{b}) > 2t - d(\mathbf{a}, \mathbf{a}') \geq 2t - t = t \geq d(\mathbf{a}, \mathbf{a}')$$

which means that C is a t error correcting code.

The proof of the converse is more complicated (Adámek [11], p/49), but the proposition has a simple geometric interpretation. Every code word of C can be thought of as a point in the n-dimensional vector space. Hence, every code word of Hamming distance of t or less would lie within a sphere centered at the code word with a radius of t. Hence, $d(C) > 2t$ implies that none of these spheres intersect. Any received vector **a'** of **a** within a specific sphere will be closed to its center **a** and thus decodable correctly, being the nearest neighbor.

4. Linear block codes: generator matrix

As one is dealing here with only two numbers 0 and 1, the arithmetic over the two bits have to be redefined. For this purpose, *congruence* of two numbers from the Theory of Numbers is employed. For a given integer $n > 1$, called *modulus*, two integers a and b are said to be *congruent modulo m* if $(a - b)/m$ = an integer and one writes a $\cong b$ (mod m). Thus, for $m = 2$

$$2 \cong 0\,(mod\,2), \quad 3 \cong 1\,(mod 2), \quad 4 \cong 0\,(mod 2), \quad 5 \cong 1\,(mod 2), \; etc. \qquad (5)$$

Thus, addition and multiplication of the bits 0 and 1 adopting modulo 2 congruence is conveniently represented in tabular form as:

+	0	1
0	0	1
1	1	0

·	0	1
0	0	0
1	0	1

This particular arithmetic is very useful in the development of very useful codes. One is by use of matrices. In the table for addition, it is noteworthy that $1 + 1 = 0$ so that $-1 = 1$.

A message of length k in a binary (n, k) block C can be formed by permutation of 0 and 1 in 2^k ways. In practice k is large, and so the dictionary of words becomes very large. For abbreviation let it be assumed that the codewords belong to k-dimensional linear vector space of n-vectors. The k basis vectors of the n-vectors can then be employed to write any code word of C by a linear combination of the basis vectors. This means that if $\mathbf{e_1}, \mathbf{e_2}, \cdots, \mathbf{e_k}$ are the (unit) basis vectors, then every code word **a** can be written as a linear combination

$$\mathbf{a} = \sum_{i=1}^{k} c_i \mathbf{e_i} \qquad (6)$$

for a unique k-tuple of scalars c_i. In other words, $c_1c_2\cdots c_k$ determine a unique code word. Eq. (6) can be written in matrix notation as

$$\mathbf{a} = \mathbf{c} \cdot \mathbf{G} \tag{7}$$

where

$$\mathbf{G} = \left[\mathbf{e_1}\ \mathbf{e_2}\cdots\mathbf{e_k}\right]^T \tag{8}$$

G is called the *generator matrix* of the code C. Evidently it is much more convenient to store **G** in the memory for generating any code word. For example, Adámek [11], p.72, Kythe and Kythe [4], p.76) consider the Hamming (7, 4) of $2^4 = 16$ code words as under:

Code word	Code word
0000 000	0110 011
1000 011	0101 010
0100 101	0011 001
0010 110	1110 000
0001 111	1101 001
1100 110	1011 010
1010 101	0111 100
1001 100	1111 111

employed by many authors for illustrative purposes of the coding methods, as the actual code words in practice are very long for unveiling the special features of the different methods of coding. The generator matrix of the code is

$$\mathbf{G} = \begin{bmatrix} 1 & 0 & 0 & 0 & 0 & 1 & 1 \\ 0 & 1 & 0 & 0 & 1 & ,0 & 1 \\ 0 & 0 & 1 & 0 & 1 & 1 & 0 \\ 0 & 0 & 0 & 1 & 1 & 1 & 1 \end{bmatrix} \tag{9}$$

Every code word of the code can be generated from it by a linear combination of the rows. For instance, the second code word is generated by taking $c_1 = 1, c_2 = c_3 = c_4 = 0$ and the third by taking $c_1 = 0, c_2 = 1, c_3 = c_4 = 0$, etc.

By a change of basis the generator matrix changes. The most systematic way is to send the message $c_0c_1\cdots c_{k-1}$ by sending it as $c_0c_1\cdots c_{k-1}d_k\cdots d_{n-1}$ whose generator matrix is

$$\mathbf{G} = [\mathbf{I}|\mathbf{P}] \tag{10}$$

where **I** is the $k \times k$ identity (unit) matrix, and **P** is a $k \times (n-k)$ matrix called the *parity matrix*. In the above example, the generator matrix is in the systematic form with

$$\mathbf{P^T} = \begin{bmatrix} 0 & 1 & 1 & 1 \\ 1 & 0 & 1 & 1 \\ 1 & 1 & 0 & 1 \end{bmatrix} \tag{11}$$

Definition. An (n, k) code C with a generator $\mathbf{G} = [\mathbf{I}|\mathbf{P}]$ in systematic form is defined to have a *parity check matrix* $\mathbf{H} = [-\mathbf{P}|\mathbf{I}']$ where I′ is the identity matrix of dimension $n - k$.

From the above definition, it immediately follows that.

Proposition 3. The matrix $\mathbf{H}^{\mathrm{T}}$ is orthogonal to $\mathbf{G}$.

Proof. $\mathbf{GH}^{\mathrm{T}} = [\mathbf{I}|\mathbf{P}]\begin{bmatrix} -\mathbf{P} \\ , \\ \mathbf{I} \end{bmatrix} = -\mathbf{P} + \mathbf{P} = 0$

This relation can be checked for the Hamming $(7, 4)$ code keeping in mind the modulo-2 arithmetic for which $1 + 1 = 0$. This means that $\mathbf{G}$ is a correct generator of the $(7, 4)$ code.

If an incorrect code word is transmitted which does not make $\mathbf{G}$ satisfy the condition $\mathbf{GH}^{\mathrm{T}} = 0$, then that word does not belong to C, and the error is detected. In this way, the parity matrix helps detect errors.

In general, one has the *Hamming codes* [17] having the properties:

$$\begin{aligned} &\text{Block Length}: && n = 2^m - 1 \\ &\text{Message Length}: && k = 2^m - m - 1 \end{aligned}$$

then it can be shown that the minimum distance of such codes is 3, and therefore, a Hamming code corrects a single error according to Proposition 2.

5. Finite fields

The modulo arithmetic over the elements of a finite set is particularly useful for algebraically extending further development of codes by introducing the definition of *fields*. As is well known, a *field* F is a set of elements $\{0, 1, a, b, c, \cdots\}$ over which two *closed operations* '$+$' (addition) and '$\cdot$' (multiplication) can be applied which satisfy the commutative, associative, and the distributive laws. For a nonzero element a, it is also assumed that a multiplicative inverse $a^{-1} \in F$exists such that $a \cdot a^{-1} = 1$.. Usually one write $a \cdot b$ simply as ab. If in a set F the multiplicative inverse does not exist, then it is called a *Ring*. If the number of elements of F is finite, it is called a *finite field*. It is easy to verify that the field $F(2)$ of the bit set $\{0, 1\}$ satisfying the mod 2 arithmetic is a finite field. However the set of all decimal integers $Z = \{0, \pm 1, \pm 2, \cdots\}$ forms an infinite *integer ring*. Similarly, the set of all polynomials $F(x) = \{a_0 + a_1 x + \cdots + a_n x^n : a_i \in F, n \geq 0\}$ also forms a polynomial ring.

Proposition 4. For every (decimal) prime p, Z_p is a field.

Proof. Since Z_p is a ring as noted earlier for Z it is only required to prove that every element $i = 1, 2, \cdots, p - 1$ possesses an inverse. Firstly, $i = 1$ has an inverse element $i^{-1} = 1$. Secondly, suppose that for $i > 1$ all the inverse elements $1^{-1}, 2^{-1}, \cdots, (i - 1)^{-1}$ exist. Now perform the integer division p/i, denoting the quotient by $q \neq 0$ and the remainder by r; then one has

$$p = qi + r \tag{12}$$

Now, as p is next to the last element $p - 1 \in Z_p$, so that in modulo p arithmetic $p \cong 0$, and Eq. (13) means that

$$-r = q \cdot i = i \cdot q \tag{13}$$

Since $q \neq 0$ and i lie between 2 and $p - 1$, it follows that $r \neq 0$ so that r^{-1} exists and $i = -q^{-1} \cdot r$. It follows that

$$i \cdot \left(-q \cdot r^{-1}\right) = -(i \cdot q) \cdot r^{-1} = -(-r) \cdot r^{-1} = 1 \tag{14}$$

which means that i^{-1} exists and is equal to $-q \cdot r^{-1}$, and the proposition is proved by induction.

The above proposition shows that a number of finite fields exist, viz. $Z_2, Z_3, Z_5, Z_7, \cdots$. Of particular interest here is the field Z_2 which will be written as $GF(2)$ and its extension $GF(2^m)$ over the elements $\left\{0, 1, \alpha, \alpha^2, \cdots, \alpha^{2^m} - 1\right\}$ which is called the Galois field named after Éveriste Galois (CE 1811 – 1832), who earned fame in his teen age to die early in a gun dual to pass into the history of stellar mathematicians. The extension is based on the treatment of polynomials over the binary field.

6. Binary field polynomials

Consider calculation with polynomials whose coefficients are the binary bits $\{0, 1\}$ of $GF(2)$. A *polynomial* $f(x)$ with one *variable* x and binary coefficients is of the form

$$f(x) = a_0 + a_1x + a_2x^2 + \cdots + a_nx^n \tag{15}$$

where $a_i == 0$ or 1 for $0 \leq i \leq n$. If $a_n = 1$, the polynomial is called *monic* of *degree* n. The variable is kept *indeterminate* and has only a formal algebraic role in coding.The polynomials of degree 1 over $GF(2)$ are evidently x and $1 + x$. Similarly, the polynomials of degree 2 are x^2, $1 + x^2$, $x + x^2$, and $1 + x + x^2$ and so on. In general, there are 2^n polynomials over $GF(2)$ with degree n.

Polynomials over Galois field $GF(2)$ can be added (or subtracted), multiplied, and divided in the usual way of treating real and complex valued polynomials.. Let

$$g(x) = b_0 + b_1x + b_2x^2 + \cdots, b_mx^m \tag{16}$$

be another polynomial over $GF(2)$, where $m \leq n$. Then, $f(x)$ and $g(x)$ are added by simply adding the coefficients of the same power of x in $f(x)$ and $g(x)$:

$$f(x) + g(x) = (a_0 + b_0) + (a_1 + b_1)x + \cdots + (a_m + b_m)x^m + a_{m+1}x^{m+1} + \cdots a_nx^n \tag{17}$$

where $a_i + b_i$ is carried out in modulo-2 addition. For example, let $\phi(x) = 1 + x + x^3 + x^5$ and $\psi(x) = 1 + x + x^2 + x^3 + x^4 + x^6$, then

$$\phi(x) + \psi(x) = (1 + 1) + x + x^2 + (1 + 1)x^3 + x^4 + x^5 + x^6 = x + x^2 + x^4 + x^5 + x^6 \tag{18}$$

as $1 + 1 = 0$. Similarly, when $f(x)$ is multiplied to $g(x)$, the following product is obtained

$$\begin{aligned}
c_0 &= a_0 b_0 \\
c_1 &= a_0 b_1 + a_1 b_0 \\
c_2 &= a_0 b_2 + a_1 b_1 + a_2 b_0 \\
&\vdots \\
c_i &= a_0 b_i + a_1 b_{i-1} + a_2 b_{i-2} + \cdots + a_i b_0 \\
&\vdots \\
c_{n+m} &= a_n b_m
\end{aligned} \tag{19}$$

the multiplication and addition of the coefficients being in modulo-2. As a result, the polynomials $\phi(x)$ and $\psi(x)$ satisfy the commutative, associative, and the distributive properties of addition and multiplication.

The division of $f(x)$ by $g(x)$ of nonzero degree can also be defined in the usual way by the *division algorithm*, so that $f(x)/g(x)$ leaves a unique quotient $q(x)$ and a remainder $r(x)$ in $GF(2)$, that is

$$f(x) = q(x)g(x) + r(x) \tag{20}$$

As an example consider $\psi(x)/\phi(x)$:

$$\begin{array}{r}
x^5 + x^3 + x + 1)\,x^6 + x^4 + x^3 + x^2 + 1\,(x \\
\underline{x^6 + x^4 \qquad + x^2 + x} \\
x^3 + x + 1
\end{array}$$

noting that $-x = x$ in $GF(2)$. Hence, $q(x) = x$ and $r(x) = 1 + x + x^3$. It can be easily verified that

$$x^6 + x^4 + x^3 + x^2 + 1 = x\left(x^5 + x^3 + x + 1\right) + x^3 + x + 1$$

If a polynomial $f(x)$ in $GF(2)$ vanishes for some value $a \in (0,1)$, then a is called a *zero* of $f(x)$ as in the case of real and complex variables. If $f(x)$ has even number of terms such as in the case of the polynomial $\phi(x)$, then it is exactly divisible by $x + 1$ as $\phi(1) = 1 + 1 + 1 + 1 = 0$. A polynomial $p(x)$ over $GF(2)$ of degree m is called *irreducible* over $GF(2)$ if $p(x)$ is not divisible by any polynomial over $GF(2)$ of degree less than m. Among the four polynomials of degree 2, viz. x^2, $x^2 + x$, $x^2 + 1$, $x^2 + x + 1$, the first three are reducible, since they are divisible by x or $x + 1$. However, $x^2 + x + 1$ does not have $x = 0$ or 1 as zero and so is not divisible by x or $x + 1$. Thus, $x^2 + x + 1$ is an irreducible polynomial of degree 2. Similarly, $x^3 + x + 1$ and $x^4 + x + 1$ are irreducible polynomials of degree 3 and 4, respectively, over $GF(2)$. It can be proved in general that (Gathen and Gerhard [9]):

Proposition 5. The polynomial $x^{2^m-1} + 1$ over $GF(2)$ is the product of all irreducible monic polynomials.

Example 1. It can be verified by multiplication of the factors on the right hand sides that.

If $m = 2$, $x^3 + 1 = (x^2 + x + 1)(x + 1)$.

If $m = 3$, $x^7 + 1 = (x^3 + x + 1)(x^3 + x^2 + 1)(x + 1)$.

If $m = 4$, $x^{15} + 1 = (x^4 + x + 1)(x^4 + x^3 + 1)(x^4 + x^3 + x^2 + x + 1)(x + 1)$, etc.

An irreducible polynomial $p(x)$ of degree m is said to be *primitive* if m is the smallest positive integer for which $p(x)$ divides $x^{2^m-1} + 1$. It is not easy to identify a primitive polynomial because of difficulty in factorizing $x^{2^m-1}+1$. However, comprehensive tables have been prepared for them for reference purposes (Lin and Costello [9]). A very short table is presented below:

m	Primitive Polynomial
2	$1+x+x^2$
3	$1+x+x^3$
4	$1+x+x^4$
5	$1+x^2+x^5$
6	$1+x+x^6$
7	$1+x^3+x^7$
8	$1+x^2+x^3+x^4+x^8$
9	$1+x^4+x^9$
10	$1+x^3+x^{10}$

Finally, the following interesting property holds.

Proposition 6. For any $l\geq 0$, $[f(x)]^{2^l}=f\left(x^{2^l}\right)$.

Proof.

$$\begin{aligned} f^2(x) &= \left[a_0+\left(a_1x+a_2x^2+\cdots+a_nx^n\right)\right]^2 \\ &= a_0^2+a_0\cdot\left(a_1x+a_2x^2+\cdots a_nx^n\right)+a_0\cdot\left(a_1x+a_2x^2+\cdots a_nx^n\right) \\ &\qquad +\left(a_1x+a_2x^2+\cdots+a_nx^n\right)^2 \\ &= a_0^2+\left(a_1x+a_2x^2+\cdots+a_nx^n\right)^2 \end{aligned}$$

as $1+1=0$. By similar repeated expansion, it follows that

$$f^2(x)=a_0^2+(a_1x)^2+\left(a_2x^2\right)^2+\cdots(a_nx^n)^2$$

Now since, $a_i=0$ or 1, $a_i^2=a_i$, so that

$$f^2(x)=a_0+a_1x^2+a_2\left(x^2\right)^2+\cdots+a_n(x^n)^2=f\left(x^2\right)$$

Squaring again one has $f^4(x)=f(x^4)$ and so on. Hence, the result for $l\geq 0$.

Corollary. If for an element $\beta, f(\beta)=0$, then $f\left(\beta^{2^l}\right)=0$. The element β^{2^l} is called a *conjugate* of β.

6.1 Construction of Galois field GF(2^m)

The Galois field extension $GF(2^m)$ from that of $GF(2)=\{0,1\}$ is obtained by introducing a new element say α, in addition to 0 and 1. Then, by definition of multiplication

$$0 \cdot \alpha = \alpha \cdot 0 = 0, \ 1 \cdot \alpha = \alpha \cdot 1 = \alpha;$$
$$\alpha^2 = \alpha \cdot \alpha, \ \alpha^3 = \alpha \cdot \alpha \cdot \alpha, \cdots, \alpha^j = \alpha \cdot \alpha \cdots \alpha \, j \, (\text{times}) \tag{21}$$

Thus, a set F is created by "multiplication" as

$$F = \{0, 1, \alpha, \alpha^2, \cdots, \alpha^j, \cdots\} \tag{22}$$

Next, a condition is put on α so that F contains only 2^m elements and is closed under the multiplication "$\cdot$". For this purpose, let $p(x)$ be a primitive polynomial of degree m over $GF(2)$ such that α is a zero of $p(x)$, that is, $p(\alpha) = 0$ over $GF(2^m)$. Since $p(x)$ divides $x^{2^m-1} + 1$ exactly according to Proposition 5, one has

$$x^{2^m-1} + 1 = q(x)p(x) \tag{23}$$

where $q(x)$ is the quotient and zero the remainder. Hence, taking $x = \alpha$,

$$\alpha^{2^m-1} + 1 = q(\alpha)p(\alpha) = q(\alpha) \cdot 0 = 0 \tag{24}$$

So that by modulo-2 addition

$$\alpha^{2^m-1} = 1 \tag{25}$$

terminating the sequence in Eq. (26) at α^{2^m-1}. Thus, under the condition $p(\alpha) = 0$, the set becomes a finite set F^* consisting of the 2^m elements

$$F^* = \{0, 1, \alpha, \alpha^2, \cdots, \alpha^{2^m-1}\} \tag{26}$$

The nonzero elements of F^* are closed under the operation of multiplication. For proving this property, consider the product $\alpha^i \cdot \alpha^j = \alpha^{i+j}$. If $i + j < 2^m - 1, \alpha^{i+j} \in F^*$. If $i + j \geq 2^m - 1$, then writing $i + j = (2^{m-1} - 1) + r$, where $0 \leq r < 2^m - 1$,

$$\alpha^{i+j} = \alpha^{2^m-1+r} = 1 \cdot \alpha^r = \alpha^r \tag{27}$$

by Eq. (26), in which α^r is an element of F^*.

The nonzero elements of F^* are also closed under the operation of addition. For this purpose, divide x^i, $(0 \leq i \leq 2^m - 1)$ by $p(x)$ the primitive polynomial of degree n, where $p(\alpha) = 0$ as before; then for $0 \leq i < 2^m - 1$ one can write

$$x^i = q_i(x)p(x) + r_i(x) \tag{28}$$

in which $q_i(x)$ and $r_i(x)$ are quotient and remainder, respectively. The remainder $r_i(x)$ is a polynomial of degree $m - 1$ or less over $GF(2)$ and hence is of the form

$$r_i(x) = r_{i0} + r_{i1}x + r_{i2}x^2 + \cdots + r_{i,m-1}x^{m-1} \tag{29}$$

Hence, setting $x = \alpha$

$$\alpha^i = r_{i0} + r_{i1}\alpha + r_{i2}\alpha^2 + \cdots + r_{i,m-1}\alpha^{m-1} \tag{30}$$

Similarly if $0 \leq j < 2^{m-1}$, α^j can be represented as a polynomial at most of degree $m - 1$. Thus, $\alpha^i + \alpha^j$ would be a polynomial at most of degree $m - 1$, and by addition of the two polynomials, the summand would be equal to some α^k where $0 \leq k < 2^{m-1}$. This proves the assertion.

Example 2. As in Lin and Costello [12], p. 32, let $m = 4$, so that $GF(2^4) = \{0, 1, \alpha, \alpha^2, \alpha^3, \alpha^4, \alpha^5, \alpha^6, \alpha^7, \alpha^8, \alpha^9, \alpha^{10}, \alpha^{11}, \alpha^{12}, \alpha^{13}, \alpha^{14}\}$. Its primitive polynomial over $GF(2)$ is $p(x) = 1 + x + x^4$, so that set $p(\alpha) = 1 + \alpha + +\alpha^4 = 0$, or adding $1 + \alpha$ to the two sides of the equation $\alpha^4 = 1 + \alpha$. Hence,

$$\alpha^5 = \alpha \cdot \alpha^4 = \alpha \cdot (1 + \alpha) = \alpha + \alpha^2, \alpha^6 = \alpha \cdot (\alpha + \alpha^2) = \alpha^2 + \alpha^3,$$
$$\alpha^7 = \alpha^3 + \alpha^4 = 1 + \alpha + \alpha^3, \alpha^8 = \alpha + \alpha^2 + \alpha^4 = 1 + \alpha^2, \ etc.$$

The highest power of α is 3 in these elements and the 4-tubles form the block code words of length 4, which can be represented in the hexadecimal code as well:

Power representation	Polynomial representation	4-tuple representation	Hexadecimal
0	0	(0 0 0 0)	0
1	1	(1 0 0 0)	8
α	α	(0 1 0 0)	4
α^2	α^2	(0 0 1 0)	2
α^3	α^3	(0 0 0 1)	1
α^4	$1+\alpha$	(1 1 0 0)	C
α^5	$\alpha + \alpha^2$	(0 1 1 0)	6
α^6	$\alpha^2 + \alpha^3$	(0 0 1 1)	3
α^7	$1 + \alpha + \alpha^3$	(1 1 0 1)	D
α^8	$1 + \alpha^2$	(1 0 1 0)	A
α^9	$\alpha + \alpha^3$	(0 1 0 1)	5
α^{10}	$1 + \alpha + \alpha^2$	(1 1 1 0)	E
α^{11}	$\alpha + \alpha^2 + \alpha^3$	(0 1 1 1)	7
α^{12}	$1 + \alpha + \alpha^2 + \alpha^3$	(1 1 1 1)	F
α^{13}	$1 + \alpha^2 + \alpha^3$	(1 0 1 1)	B
α^{14}	$1 + \alpha^3$	(1 0 0 1)	9

Proposition 7. The elements of $GF(2^m)$ form all the zeros of $x^{2^m} + x$.

Proof. The proposition obviously holds for the element 0. For a nonzero element $\beta \in GF(2^m)$ let, $\beta = \alpha^i$ then $\beta^{2^m - 1} = \alpha^{i(2^m - 1)} = \left(\alpha^{2^m - 1}\right)^i = 1^i = 1$ by Eq. (26). Hence, by modulo-2 addition one has $\beta^{2^m - 1} + 1 = 0$ or, $\beta^{2^m} + \beta = 0$.

The next proposition determines the minimal polynomial corresponding to an element $\beta \in GF(2^m)$:

Proposition 8. If f e is the smallest integer for which $\beta^{2^e} = \beta$, then the minimal polynomial $\phi(x)$ corresponding to $\beta \in GF(2^m)$ is given by

$$\phi(x) = \Pi_{i=0}^{e-1}\left(x + \beta^{2^i}\right) \tag{31}$$

Proof. Since β is a zero of $\phi(x)$, $\phi(\beta) = 0$. Hence, by the Corollary to Proposition 6, $\phi\left(\beta^{2^i}\right) = 0$ which means that β^{2^i} is also a zero of $\phi(x)$. Hence, $\phi(x)$ is the product of the factors $x + \beta^{2^i}$, provided that i is one less than that makes $\beta^{2^i-1} = 1$ or, $\beta^{2^i} = \beta$.

Example 3. For $\beta = \alpha^3 \in GF(2^4)$, the conjugates of β are $\beta^2 = \alpha^6, \beta^{2^2} = \alpha^{12}, \beta^{2^3} = \alpha^{24} = \alpha^9$. The minimal polynomial of β is $\phi(x) = (x + \alpha^3)(x + \alpha^6)(x + \alpha^{12}(x + \alpha^9) = (x^2 + \alpha^2 x + \alpha + \alpha^3)[x^2 + (1 + \alpha^2)x + \alpha^2 + \alpha^3] = x^4 + x^3 + x + 1$, using the table given above, with $\alpha^{15} = 1$.The following table gives the minimal polynomials for all the powers of α (Adámek [11], p. 216):

Elements	Minimal Polynomial
0	x
1	x + 1
$\alpha, \alpha^2, \alpha^4, \alpha^8$	$x^4 + x + 1$
$\alpha^3, \alpha^6, \alpha^9, \alpha^{12}$	$x^4 + x^3 + x^2 + 1$
α^5, α^{10}	$x^2 + x + 1$
$\alpha^7, \alpha^{11}, \alpha^{13}, \alpha^{14}$	$x^4 + x^3 + 1$

Thus, all the irreducible minimal polynomial factors of $x^{2^4-1} + 1$ are obtained.

7. Cyclic codes

Cyclic codes introduced by Prange [13] form an important subclass of linear codes by the introduction of additional algebraic structure that a cyclic shift of a code word is also a code word. Thus, if in a code word $\mathbf{a} = (a_0, a_1, \cdots, a_{n-1})$ of a cyclic code, a shift of one place to the right is made a new word $\mathbf{a}^{(1)} = (a_{n-1}, a_0, a_1, \cdots, a_{n-2})$ is formed. Similarly, a shift of l places to the right is made, then the word

$$\mathbf{a}^{(l)} = (a_{n-l}, a_{n-l+1}, \cdots, a_{n-1}, a_0, a_1, \cdots, a_{n-l-1}) \tag{32}$$

is formed. It may be checked that the (7, 4) Hamming code noted after Eq. (8) is not a cyclic code.

The special algebraic properties of the cyclic codes are described by a polynomial representation formed by the component elements of **a**:

$$f(x) = a_0 + a_1 x + a_2 x^2 + \cdots + a_{n-1} x^{n-1} \tag{33}$$

where x is *indeterminate,* and the degree is $n - 1$ or less according to $a_{n-1} = 1$ or 0. Here, the polynomial is called a *code polynomial*, and the code word $\mathbf{a}^l$ is represented by the polynomial

$$f^{(l)}(x) = \left(a_{n-l} + a_{n-l+1}x + \cdots + a_{n-1}x^{l-1}\right) + \left(a_0 x^l + a_1 x^{l+1} + \cdots + a_{n-l-1}x^{n-1}\right) \tag{34}$$

The polynomial $f^{(l)}(x)$ can be represented in a compact form.

Proposition 9. $f^{(l)}(x) = x^l f(x) \qquad (mod\, x^n + 1).$

Proof.

$$\begin{aligned} x^l f(x) &= a_0 x^l + a_1 x^{l+1} + \cdots + a_{n-l-1} x^{n-1} + a_{n-l} x^n + \cdots + a_{n-1} x^{n+l-1} \\ &= \left(a_{n-l} + a_{n-l+1} x + \cdots + a_{n-1} x^{l-1}\right) + a_0 x^l + \cdots + a_{n-l-1} x^{n-1} \\ &+ a_{n-l}(x^n + 1) + a_{n-l+1} x(x^n + 1) + \cdots + a_{n-1} x^{l-1}(x^n + 1) \\ &= q(x)(x^n + 1) + f^{(l)}(x) \end{aligned} \tag{35}$$

where $q((x) = a_{n-l} + a_{n-l+1}x + \cdots + a_{n-1}x^{l-1}$, which proves the theorem.

In general, consider an (n, k) cyclic code C that possesses a code polynomial

$$g(x) = 1 + g_1 x + g_2 x + \cdots + g_{r-1} x^{r-1} + x^r \tag{36}$$

of minimum degree r (such as $1 + x + x^3$ in the above given example), then the polynomials $xg(x) = g^{(1)}(x), x^2 g(x) = g^{(2)}(x), \cdots, x^{n-r-1} g(x) = g^{(n-r-1)}(x)$ represent cyclic shifts of the code polynomial $g(x)$ that are code polynomials in C. Since C is linear, a linear combination

$$\begin{aligned} \psi(x) &= b_0 g(x) + b_1 x g(x) + \cdots + b_{n-r-1} x^{n-r-1} g(x) \\ &= \left(b_0 + b_1 x + \cdots + b_{n-r-1} x^{n-r-1}\right) g(x) \end{aligned} \tag{37}$$

is also a code polynomial with $b_i = 0$ or 1 of degree $n - 1$ or less. The number of such polynomials is 2^{n-r}. However, there are 2^k code polynomials in C, so that $n - r = k$ or, $r = n - k$. Thus,

Proposition 10. In an (n, k) cyclic code C, one unique code polynomial

$$g(x) = 1 + g_1 x + g_2 x^2 + \cdots + g_{n-k-1} x^{n-k-1} + x^{n-k} \tag{38}$$

of degree $n - k$ called the *generator polynomial* yields every binary polynomial code of degree $n - 1$ or less by multiplication with another polynomial of degree $k - 1$.

From Eq. (39), one can find a generator polynomial from the following:

Proposition 11. The generator polynomial $g(x)$ of an (n, k) cyclic code is a factor of $x^n + 1$.

Proof. Multiplying $g(x)$ by x^k in Eq.(39), one obtains a polynomial $x^k g(x)$ of degree n. Hence, according to Eq. (36)

$$x^k g(x) = (x^n + 1) + g^{(k)}(x) \tag{39}$$

where $g^k(x)$ is the remainder, which is a polynomial obtained by k cyclic shifts of the coefficients of $g(x)$. Hence, $g^{(k)}(x)$ is a multiple of $g(x)$, say $\psi(x)$ of degree k, viz. $g^{(k)}(x) = \psi(x) g(x)$. Thus,

$$x^n + 1 = \{x^k + \psi(x)\} g(x) \tag{40}$$

which completes the proof.

In practice where n is large $x^n + 1$ may have several factors of degree $n - k$ to describe cyclic codes. For example (Lin and Costello [12], p. 90)

$$x^7 + 1 = (1 + x)(1 + x + x^3)(1 + x^2 + x^3)$$

in which either of the last two factors can be selected for a generator polynomial. In the example considered earlier, we selected $g(x) = 1 + x + x^3$. In practice, it is difficult to make a choice because of implementation difficulty. Nevertheless it is possible to form the generator matrix of a cyclic code and parity-check matrix for error detection and correction. Lin and Costello [9] give an excellent treatment of the whole topic.

8. The BCH codes

The Bose, Chaudhuri, Hocqenghem (BCH) [14, 15] codes form a large class of powerful random error-correcting codes (Lin and Costello [9]). It generalizes the Hamming code in the following manner:

$$\begin{array}{ll} \text{Block length :} & n = 2^m - 1 \\ \text{Parity check bits :} & n - k \leq mt \\ \text{Minimum distance :} & d \geq 2t + 1 \end{array} \tag{41}$$

where $m \geq 3$, and $t < 2^{m-1}$.

This code is capable of correcting any combination of t or fewer errors in a code word of length $n = 2^m - 1$ bits and is called a t−error−correcting code.The generator polynomial of the code $g(x)$ of length $2^m - 1$ is the *lowest degree polynomial* over $GF(2)$ which has

$$\alpha, \alpha^2, \alpha^3, \cdots, \alpha^{2t} \tag{42}$$

as its zeros, that is to say, $g(\alpha^i) = 0$ for $1 \leq i \leq 2t$. It follows from the Corollary to Proposition 6 that the conjugates of the zeros $\alpha, \alpha^2, \alpha^3, \cdots, \alpha^{2t}$ are also zeros of $g(x)$. If $\phi_i(x)$ is the *minimal polynomial* of α^i, then $g(x)$ must be the *lowest common multiple* (LCM) of $\phi_1(x), \phi_2(x), \cdots \phi_{2t}(x)$, that is

$$g(x) = \text{LCM}\{\phi_1(x), \phi_2(x), \cdots, \phi_{2t}(x)\} \tag{43}$$

Now, if i is an even integer, it can be written as $i = i' 2^l$ where i' is an odd number and $l \geq 1$. Then, for such i, $\alpha^i = \left(\alpha^{i'}\right)^{2^l}$, which is conjugate of $\alpha^{i'}$, and therefore, α^i and $\alpha^{i'}$ have the same minimal polynomial, that is

$$\phi_i(x) = \phi_{i'}(x) \tag{44}$$

Hence, every even power of α in the sequence (44) has the same minimal polynomial as some preceding odd power of α in the sequence. As a result, the generator $g(x)$ given by Eq. (44) reduces to

$$g(x) = \text{LCM}\{\phi_1(x), \phi_3(x), \cdots, \phi_{2t-1}(x)\} \tag{45}$$

Since the degree of each minimal polynomial is m or less, the degree of $g(x)$ is at most mt. This means that the number of parity-check bits $n - k$ of the code is at most

mt. There is, however, no simple formula for enumerating the parity $n-k$ bits, but tables do exist for various values of $n, k,$ and t (Lin and Costello [9]).

In the special case of single error correction for a $n = 2^m - 1$ long code word, $g(x) = \phi_1(x)$ which is a primitive polynomial of degree 2^m. Thus, the single error correcting BCH code of length $2^m - 1$ is a cyclic Hamming code.

Example 4. As in Lin and Costello [12], p. 149, let α be a primitive element of $GF(2^4)$ such that $\alpha^4 + \alpha + 1 = 0$. From Example 2,

$$\begin{aligned} \phi_1(x) &= 1 + x + x^4 \\ \phi_3(x) &= 1 + x + x^2 + x^3 + x^4 \end{aligned}$$

Hence, the *double error* correcting BCH code of length $n = 2^4 - 1 = 15$ is generated by

$$\begin{aligned} g(x) &= \mathrm{LCM}\{\phi_1(x), \phi_3(x)\} = \phi_1(x)\phi_3(x) \\ &= (1 + x + x^4)(1 + x + x^2 + x^3 + x^4) \\ &= 1 + x^4 + x^6 + x^7 + x^8 \end{aligned}$$

It is possible to construct the parity matrix H of this code and is presented in Lin and Costello [9].

9. The RS codes

In the construction of BCH codes, the generator polynomials over $GF(2)$ were considered using the elements of the minimal polynomials over the extended field $GF(2^m)$. Since the minimal polynomial for an element β also has all the conjugates of β as the zeros of the minimal polynomial, the product of the minimal polynomials usually exceeds the number $2t$ of the specified zeros. In the Reed–Solomon (RS) codes [16], this situation is tackled by considering the extended $GF(2^m)$ as the starting point. An element $\beta \in GF(2^m)$ has obviously the minimal polynomial $x + \beta$. If $\beta = \alpha^i$ where $\alpha \in GF(2^m)$, the required generator $g(x)$ of degree $2t$ as in Eq. (46) becomes

$$g(x) = (x + \alpha^i)(x + \alpha^{i+1}) \cdots (x + \alpha^{i+2t-1}) \tag{46}$$

There are no extra zeros in $g(x)$ included by the conjugates of the minimal polynomial. So the degree of $g(x)$ is exactly $2t$. Thus, $n - k = 2t$ for a RS code and the minimum distance by Proposition 2 is $d = 2t + 1 = n - k + 1$. Hence, an RS code is

$$\begin{aligned} &\text{Block length}: && n = 2^m - 1 \\ &\text{Parity check bits}: && n - k = 2t \\ &\text{Minimum distance}: && d = n - k + 1 \end{aligned} \tag{47}$$

Example 5. Let $n = 2^4 - 1 = 15$, and consider three error codes $(t = 3)$ over $GF(2^4)$, having the primitive polynomial $p(x) = 1 + x + x^4$. Taking $i = 1$ in Eq. (47) the required generator polynomial is

$$g(x) = (x + \alpha)(x + \alpha^2)(x + \alpha^3)((x + \alpha^4)(x + \alpha^5)(x + \alpha^6)$$

where $\alpha^4 = 1 + \alpha$, and $\alpha^{15} = 1$. Thus simplifying, one gets the generator polynomial for the (15, 9) code over $GF(2^4)$ as

$$g(x) = \alpha^6 + \alpha^9 x + \alpha^6 x^2 + \alpha^4 x^3 + \alpha^{14} x^4 + \alpha^{10} x^5 + x^6$$

The corresponding code word is $\alpha^6 \alpha^9 \alpha^6 \alpha^4 \alpha^{14} \alpha^{10} 1$. Using the table of Example 2, one obtains the binary equivalent over $GF(2)$ as

$$0011\,0101\,0011\,1100\,1001\,1110\,1000$$

The length of the binary word is 28. This type of code is called *derived binary code*.

10. Conclusion

In this information age transmission of data is all pervasive due to immense development of electronic technology. In this general scenario, it is rarely recognized that the foundations of this information web were layed by "A mathematical theory of communication" propounded by Claude E. Shannon [2] that called for means of error-free transmission of information as digital data. This realization led to the creation of Theories of Coding by more mathematicians, pioneered by Richard W. Hamming [17] emphasizing that data need to be appropriately coded for realizing the objectives. As digital data consist of just two bits 0 and 1, special algebra were employed which were based on the algebra of finite fields. In this endeavor, several types of data coding have been discovered and put to practical use depending on the application. Some of the codes developed depend on polynomials over the binary field of 0 and 1; prominent among them being the BCH [14, 15] and the RS codes [16] that are widely used in practice. This chapter gives a simple mathematical preview of this type of code development and is intended for those who may be interested to further delve into the subject of coding of digital data.

Acknowledgements

The author is thankful to the SN Bose National Center for Basic Sciences for supporting this research, as also to the referees for their learned comments for the improvement in the presentation of this article.

Disclosures

There is no conflict of interests in the presented chapter.

Data availability

No data were generated or analyzed in the presented chapter.

Author details

Sujit K. Bose
S.N. Bose National Center for Basic Sciences, Salt Lake City, Kolkata, West Bengal, India

*Address all correspondence to: sujitkbose1@gmail.com

References

[1] Leach DP, Malvino AP, Saha G. Digital Principles and Applications. New Delhi: Tata McGraw Hill Education; 2011

[2] Shannon CE. A mathematical theory of communication. Bell System Technology Journal. 1948;**27**:379-423

[3] Bose R. Information Theory, Coding and Cryptography. New Delhi: McGraw Hill Education; 2016

[4] Kythe DK, Kythe PK. Algebraic and Stochastic Coding Theory. Boca Raton: CRC Press; 2012

[5] Roth RM. Introduction to Coding Theory. Cambridge: Cambridge University Press; 2006

[6] Moon TK. Error Correction Coding: Mathematical Methods and Algorithms. Hoboken, New Jersey: Wiley-Interscience

[7] Ling S, Xing C. Coding Theory a First Course. Cambridge: Cambridge University Press; 2004

[8] Blahut RE. Alebraic Codes for Data Transmission. Cambridge: Cambridge University Press; 2003

[9] Gathen JW, Gerhard J. Modern Computer Algebra. Cambridge: Cambridge Universit Press; 2003

[10] Proakis JG, Salehi M. Digital Communications. Boston: McGraw Hill Higher Education; 2001

[11] Adámek J. Foundations of Coding. New York: John Wiley & Sons; 1991

[12] Lin S, Costello DJ Jr. Error Control Coding: Fundamentals and Applications. Prentice-Hall; 1983

[13] Prange E. Cyclic Error-Correcting Codes in Two Symbols, AFCRC-TN-57. Cambridge: Air Force Cambridge Research Center; 1957

[14] Bose RC, Ray-Chaudhuri DK. On a class of error correcting binary group codes. Information and Control. 1960;**3**: 68-79

[15] Hocquenghem A. Codes Correcteurs d'erreurs. Chiffres. 1959;**2**:147-156

[16] Reed IS, Solomon G. Polynomial codes over certain finite fields. Journal of the Society for Industrial and Applied Mathematics. 1960;**8**:300-304

[17] Hamming RW. Error detecting and error correcting codes. Bell System Technical Journal. 1950;**29**:147-160

Chapter 3

Orthogonal Polynomials Based Operational Matrices with Applications to Bagley-Torvik Fractional Derivative Differential Equations

Imran Talib and Faruk Özger

Abstract

Orthogonal polynomials are the natural way to express the elements of the inner product spaces as an infinite sum of orthonormal basis sets. The construction and development of the many important numerical algorithms are based on the operational matrices of orthogonal polynomials including spectral tau, spectral collocation, and operational matrices approach are few of them. The widely used orthogonal polynomials are Legendre, Jacobi, and Chebyshev. However, only a few papers are available where the Hermite polynomials (HPs) were exploited to solve numerically the differential equations. The notable characteristic of the HPs is its ability to approximate the square-integrable functions on the entire real line. The prime objective of this chapter is to introduce the two new generalized operational matrices of HPs which are developed in the sense of the Riemann-Liouville fractional-order integral operator and Hilfer fractional-order derivative operator. The newly derived operational matrices are further used to construct a numerical algorithm for solving the Bagley–Torvik types fractional derivative differential equations (FDDE). Moreover, the results obtained by using the proposed algorithm are compared with the results obtained otherwise to demonstrate the efficiency and accuracy of the proposed numerical algorithm. Some examples are solved for application purposes.

Keywords: operational matrices approach, Hilfer fractional derivative operator, Sylvester type matrix equations, Hermite polynomials, inner product spaces, Hilbert spaces, spectral methods, Bagley–Torvik fractional-order differential equations

1. Introduction

It's a proven fact that every vector space V generated by a finite set of vectors has a basis [1]. But what's the situation if V is not finitely generated? For example $C([0,1])$, the vector space of all real-valued continuous functions defined on the compact interval $[0,1]$; $P(\mathbb{R})$, the vector space of all real-valued polynomials defined on $\mathbb{R}$; and

$\mathbb{R}^\infty$, the vector space of infinite sequences $(\beta_1, \beta_2, \cdots,)$ of real numbers. Do these spaces have bases? If they have then how can we construct them? The case for $P(\mathbb{R})$ is so obvious because every $f \in P(\mathbb{R})$ can be expressed as a finite linear combination of the infinite set of polynomials $\{1, x, x^2, x^3, \cdots, x^n, \cdots\}$. However, the case for $\mathbb{R}^\infty$ is so surprising because there is no general way to add together infinitely many vectors in a vector space. So what about the basis of $\mathbb{R}^\infty$? Since $\mathbb{R}^\infty$ is a natural generalization of $\mathbb{R}^n$, therefore, consider the set $S = \{e_1, e_2, e_3, \cdots, e_n, \cdots\}$ which generalizes the standard basis for $\mathbb{R}^n$. Clearly, S is linearly independent but it does not span $\mathbb{R}^\infty$ because every finite linear combination of the e_i's is a vector having only finitely many components nonzero. For example, the vector $u = \{1, 1, \cdots, \} \in \mathbb{R}^\infty$ can not be expressed as linear combination of the vectors $e_i = \{0, 0, \cdots, 0, 1, 0 \cdots\}$. Therefore, the set S together with vector u must be a linearly independent set, so that if it spans $\mathbb{R}^\infty$ then it must be a basis set. But unfortunately, it does not span $\mathbb{R}^\infty$ because the vector $v = \{1, 2, 3, \cdots\}$ can not be written as a linear combination of the vectors u and e_i's. Continuing in the same way, one can enlarge the set S to a maximal linearly independent set, such that, it generates $\mathbb{R}^\infty$. But will this process eventually be terminated, and produces the basis for $\mathbb{R}^\infty$? Actually, we are unable to construct any countable set of vectors in $\mathbb{R}^\infty$ that may span it. Although, Zorn's Lemma can be applied to show that every V has bases, see [2–4]. However, it does not explain the procedure of how to actually construct these bases if V is not finitely generated. So, the construction of bases for infinite dimensional vector spaces is a serious problem that requires some alternative ideas where the sum of infinitely many vectors makes some sense.

One might claim that the vector $u \in \mathbb{R}^\infty$ can be uniquely expressed as an "*infinite linear combination*" of the elements of S, such that, $u = \sum_{j=1}^{\infty} \beta_j e_j$. But the generally infinite sum of vectors does not convey any sense due to many reasons. For instance, can we make some sense by adding the vectors $(1, 1, \cdots)$, $(2, 2, \cdots)$, and $(3, 3, \cdots)$? Even algebraic manipulations with this infinite sum will lead to some serious problems. Might be some certain infinite sums in certain vector spaces, like $\mathbb{R}^\infty$ make some sense but they can not be generalized to other settings. So, we have to come up with alternative ways where the idea of infinite sums conveys some sense in general settings.

The only technique that can provide sense to adding infinitely many vectors is to consider the sequence of partial sums, $P_k = \sum_{j=1}^{k} u_j$, $k = 1, 2, \cdots,$ provided that it converges. There the convergence means that the sequence $\{P_k\}$ is getting closer and closer to some fixed P as $k \to \infty$, i.e., the distance between $\{P_k\}$ and P is getting smaller and smaller with increasing k. Therefore, we need the notion of defining the distance in vector spaces to provide sense to an infinite sum of vectors $\sum_{j=1}^{\infty} u_j$. However, the distance can be defined as those vector spaces for which we have an inner product. Consequently, the inner product spaces are the natural way to give sense to infinite sums of vectors.

Now our question of finding the basis set for infinite dimensional vector spaces turns into finding an orthonormal set $\{u_n\}$ such that it spans whole space V. If it happens then the set $\{u_n\}$ is called the orthonormal basis for V. This idea is very useful in many applications of mathematics, particularly in approximation theory, see [5–12].

Orthogonal polynomials are the best alternative way that provides a meaningful sense of an infinite sum of vectors, provided that this sum converges to some fixed vector f in Hilbert spaces which are the complete inner product spaces. These polynomials provide an orthonormal basis that generates the L^2 spaces which are the prototypical examples of Hilbert spaces. The elements of L^2 spaces are the square-integrable functions, i.e., all functions f either defined on finite, semi-infinite, or infinite intervals must have $\int_{\text{lower limit}}^{\text{upper limit}} f^2 dx$ finite. it is worth mentioning that the L^2 spaces provide the L^2 convergence or "convergence in mean" rather than point-wise convergence. The most frequently used orthogonal polynomials are Legendre, Jacobi, Chebyshev, Laguerre, Chelyshkov, and HPs, see [13–18].

The aforementioned polynomials have widespread implications for solving a wide range of problems in Mathematics and its related disciplines. Numerous problems in science and engineering which contain differential and integral equations have been solved by using these polynomials, see [8, 10, 17–19]. The frameworks of many important numerical methods are dependent on the orthogonal polynomials, for instance, but not limited to spectral tau method, spectral collocation method, and operational matrices approach, see [7, 8, 15, 16].

Motivated by the aforementioned studies, our prime objective is to reveal the applicability of the HPs for solving the Bagley–Torvik types FDDE where the fractional-order derivatives are considered in the sense of Hilfer. These polynomials are very practical for solving those problems in which the solution is defined on the entire real line. Additionally, these polynomials have widespread applications in various areas of Physics, Economics, and Biology. For instance, in the problems of meteorology and coastal hydrodynamics, see [20]; in the problems of biological and epidemiological sciences where the HPs were employed to reduce the multi-dimensional system of ordinary differential equations into a system of algebraic equations, see [21]; in the problems of Economics, where the HPs method was used to express the behavior of financial variables, see [22]. In addition, HPs have been extensively used in the modeling of non-Gaussian excitations that reflect models of numerous phenomena surrounding us, see [23, 24].

We consider the following Bagley–Torvik types FDDE [25].

$$\begin{aligned} &\lambda_3 x''(t) + \lambda_{1\,\mathrm{H}}\mathcal{D}^{\alpha,\beta} x(t) + \lambda_2 x(t) = y(t), t \in (-\infty, \infty), \\ &x(0) = c_1, \quad x'(0) = c_2, \end{aligned} \tag{1}$$

where c_1, c_2, λ_1, λ_2, and λ_3 are arbitrary real constants with $\lambda_3 \neq 0$. The fractional-order derivative is in the sense of Hilfer, and $y(t)$ is the source term. The analytical solution of the problem (1) for $c_1 = c_2 = 0$ can be computed by solving the following integral equation

$$x(t) = \int_0^t G(t-z) y(z) dz, \tag{2}$$

where G is a green function given as under

$$G(t) = \frac{1}{\lambda_3} \sum_{l=0}^{\infty} \frac{(-1)^l}{\Gamma(l+1)} \left(\frac{\lambda_2}{\lambda_3}\right)^l t^{2l+1} E^{(l)}_{1/2,2+3l/2}\left(-\frac{\lambda_1}{\lambda_3}\sqrt{t}\right), \tag{3}$$

The expression $E^{(l)}_{\gamma,\delta}$ is the lth derivative of the two parametric Mittagg-Leffler function, given as ([26], 8.26)

$$E_{\gamma,\delta}^{(l)}(t) = \sum_{i=0}^{\infty} \frac{\Gamma(i+l+1)t^i}{\Gamma(i+1)\Gamma(\gamma i+\gamma l+\delta)}, l = 0, 1, \cdots. \tag{4}$$

The analytical solution (2) of the problem (1) involves the convolution integral that consists of Green's function which is hard to compute for the generalized functions due to the involvement of the infinite sums of the derivatives of the Mittag-Leffler function. That complication motivated the development of the numerical methods for solving (1).

In the literature, various numerical methods have been used to obtain the approximate solution of the problem (1). A few of them are listed there: in [27], the author solved (1) by introducing the exponential integrators; in [28], the authors developed the collocation–shooting technique to solve (1); in [29], the authors introduced the Taylor matrix method to approximate the solution function of (1); in [25], the authors developed the alternative numerical schemes to solve (1) by introducing its discretization which is based on fractional linear multistep methods; in [30], the author proposed the Bessel collocation method to solve numerically the problem (1); in [31], the authors approximated the solution function of (1) by using the basis of the second kind Chebyshev wavelet; in [32], the authors solved (1) by proposing an analytical technique based on the variational iteration method and the Adomian decomposition method; and in [33], the authors proposed the analytical solution of (1) by using the Adomian decomposition technique. For more study on the analytical and approximate techniques developed for obtaining the solution of the problem (1), we refer the reader to study ([26], p. 230), ([34], Thm. 4.1), cf. ([26], Eqs. (8.26) and (8.27)), [35–38].

We introduced an operational matrices approach for computing the approximate solution to the problem (1). The framework of the proposed approach is based on the fractional-order integral and fractional-order derivative operational matrices of HPs. The fractional-order derivative is considered in the sense of Hilfer. By means of the operational matrices, the problem (1) is transformed into Matrix Equations which are then solved by using the *Matlab* built-in function, *lyap*. Finally, the solution of (1) is approximated by using the basis of HPs. The proposed approach is easier to use than spectral Tau and spectral collocation methods when the solution is approximated as the basis of HPs. Because HPs provide the approximation of the solution function on the entire real line, thus involve the improper integrals to compute the series coefficients and to determine the residual functions as the case of the spectral Tau method, see ([39], Eq. (29)). So generally, it's hard to compute the improper integral for the generic functions by using the analytical techniques of integrations. We have to approximate those integrals numerically which may compromise the accuracy. However, the proposed approach is independent of computing the residual functions and the choice of suitable collocation points. Additionally, the proposed approach transforms the problems into Sylvester equations that involve an unknown vector determined by using the *Matlab* built-in function, *lyap*. The unknown vector is then used to approximate the solution functions of the problems. It's worth mentioning that we introduce the new generalized derivative operational matrix developed in the sense of Hilfer. Also, the problem (1) is not yet to be solved with Hilfer fractional-order derivatives.

2. Fractional calculus

The scholarly discussion between two great names of the nineteenth century, L'Hospital and Leibniz opened the discussion on the urge and development of noninteger-order derivatives and integral operators. For many years, the subject of Fractional calculus (FC) had been considered as an abstract mathematical idea without having applications in physics and engineering sciences. However, the notable contributions of some renowned scientists, Euler, Laplace, Fourier, Abel, Liouville, Grunwald, Letnikov, Riemann, Laurent, Heaviside, Weyl, Hardy, Riesz, Caputo, Samko, Srivastava, Oldham, Osler, Mainardi, Love, Spanier, Ross, Bagley, Torvik, Baleanu, Atangana, and Katugampola provide the wings to FC, and now it's soaring in the sky due to its immense applications in every field of sciences, see [26, 40–50].

The fractional derivative operators can not be uniquely defined like integer-order derivative operators. Scientists developed various types of fractional derivative operators to observe nature in a precise way. So expressing the physical phenomena with a single derivative can not capture their various attributes because nature is not constant, it's evolving and developing in every second. Therefore, there is a strong need for generalized operators that should have abilities to demonstrate the generic behavior of the physical phenomena, see ([51], Chap. 5,7, and [15, 52, 53]). The most commonly used fractional derivative operators are the Riemann–Liouville and Caputo expressed in the following way [26]:

$$ {}_{\mathrm{RL}}J^{\alpha}_{b^+}x(t) = \frac{1}{\Gamma(\alpha)}\int_b^t (t-y)^{\alpha-1}x(y)\mathrm{d}y, \quad t>b, \quad \alpha>0. \tag{5} $$

Thus the fractional-order derivative operators in Riemann–Liouville and Caputo senses can be expressed as

$$ \begin{aligned} \mathrm{RL}\mathcal{D}^{\alpha}_{b^+}x(t) &= D^n {}_{\mathrm{RL}}J^{n-\alpha}_{b^+}x(t) \\ &= \frac{1}{\Gamma(n-\alpha)}\frac{\mathrm{d}^n}{\mathrm{d}t^n}\int_b^t (t-y)^{n-\alpha-1}x(y)\mathrm{d}y, \quad t>b, \\ \mathrm{C}\mathcal{D}^{\alpha}_{b^+}x(t) &= {}_{\mathrm{RL}}J^{n-\alpha}_{b^+}D^n x(t) \\ &= \frac{1}{\Gamma(n-\alpha)}\int_b^t (t-y)^{n-\alpha-1}x^{(n)}(y)\mathrm{d}y, \quad t>b, \end{aligned} \tag{6} $$

respectively, where $n-1<\alpha<n, n\in\mathbb{N}$, and $\alpha>0$. The following results about Riemann–Liouville fractional-order integral and Caputo fractional-order derivative operators are very useful in computing the operational matrices, see ([7], Thm. 4.7) and ([39], Thm. 1).

$$ \begin{aligned} &{}_{\mathrm{RL}}J^{\alpha}_{b^+}(t-b)^l = \frac{\Gamma(l+1)}{\Gamma(l+1+\alpha)}(t-b)^{l+\alpha}, \\ &\text{and} \\ &{}_{\mathrm{C}}\mathcal{D}^{\alpha}_{b^+}(t-b)^l = \frac{\Gamma(l+1)}{\Gamma(l+1-\alpha)}(t-b)^{l-\alpha}, l\in\mathbb{R}_+, \& \, l\geq\lceil\alpha\rceil. \end{aligned} \tag{7} $$

Generalization is a very useful process that allows researchers to make inferences for a wide class of problems. Mathematicians are always curious about developing generalized results that allow them to recognize the similarities in results acquired in one circumstance and cover many useful results as special cases. So Hilfer proposed a generalized fractional-order derivative operator that treats (6) as special cases, see [46].

Definition 1. The generalized fractional-order derivative in Hilfer's sense is defined as

$$
\begin{aligned}
\mathrm{H}\mathcal{D}_{b^+}^{\alpha,\beta}x(t) &= \left(I_{b^+}^{\beta(n-\alpha)}\frac{d^n}{dt^n}I^{(1-\beta)(n-\alpha)}x\right)(t) \\
&= \left(I_{b^+}^{\delta-\alpha}\frac{d^n}{dt^n}I_{b^+}^{n-\delta}x\right)(t), \quad t\in[b,c],
\end{aligned} \tag{8}
$$

where $n-1<\alpha<n$, $0<\beta<1$, $\delta=n\beta+\alpha(1-\beta)$, and $\frac{d^n}{dt^n}$ is in classical sense.
Remark 2.1.

1. If $\beta=0$, then Definition 1 is the expression for Riemann–Liouville fractional-order derivative.

2. If $\beta=1$, then Definition 1 is the expression for Caputo fractional-order derivative.

Lemma 2.2. [54] *For $0<\alpha<1$, $0<\beta<1$, and $l\in\mathbb{N}$, we have the following result*

$$
{}_{\mathrm{H}}\mathcal{D}_{b^+}^{\alpha,\beta}(t-b)^l = \frac{\Gamma(l+1)}{\Gamma(l+1-\alpha)}(t-b)^{l-\alpha}, l\in\mathbb{R}_+, \&l\geq\lceil\alpha\rceil.
$$

Lemma 2.3. *If $l-2(s+r)+j+r\in\mathbb{N}$, then we have the following result*

$$
\int_{-\infty}^{\infty} t^{l-2(s+r)+j+r}\exp(-t^2)dt = \Gamma\left(\frac{l-2(s+r)+j+r+1}{2}\right). \tag{9}
$$

Proof. Substituting $t^2=u$ and integration by parts provide the required result. □

3. Hermite polynomials

HPs are the classical orthogonal polynomials that have the ability to approximate any square-integrable function on the entire real line. These polynomials have widespread applications in many areas of applied sciences, including Physics, Economics, and Biolog, see [20–24]. This section is devoted to illustrating some useful properties of HPs.

HPs can be defined using the following analytical expression, see [55].

$$
H_j(t) = \sum_{r=0}^{\lfloor\frac{j}{2}\rfloor}\frac{(-1)^r\Gamma(j+1)(2t)^{j-2r}}{\Gamma(r+1)\Gamma(j-2r+1)}, j=0,1,2,\cdots, \quad t\in(-\infty,\infty), \tag{10}
$$

where the notation $\lfloor j\rfloor$ is the floor function that takes input as a real number j and exhibits as output the greatest integer less than j. Using (10), one may compute the following HPs for $j=0,1,\cdots,4$ as

$$H_j(t) = \begin{cases} 1, & \text{for} j = 0, \\ 2t, & \text{for} j = 1, \\ 4t^2 - 2, & \text{for} j = 2, \\ 8t^3 - 12t, & \text{for} j = 3, \\ 16t^4 - 48t^2 + 12, & \text{for} j = 4. \end{cases} \tag{11}$$

The orthogonality conditions for HPs with respect to the weight function, $w(t) = exp\,(-t^2)$ are listed there

$$\int_{-\infty}^{\infty} w(t)H_j(t)H_i(t)\mathrm{d}t = \begin{cases} 0, & \text{for} i \neq j, \\ \sqrt{\pi}2^j\Gamma(j+1) & \text{for} i = j. \end{cases} \tag{12}$$

3.1 Useful properties of HPs

In this section, we list some interesting properties of HPs that are useful to construct integer-order derivative and integer-order integral operational matrices of HPs.

The following are the HPs recurrence relations, see [56].

$$H_j^{(1)}(t) = 2jH_{j-1}(t), \text{for} j \geq 1. \tag{13}$$

$$H_{j+1}(t) = 2tH_j(t) - 2jH_{j-1}(t), \text{for} j \geq 1. \tag{14}$$

Any square-integrable function, i.e., $x(t) \in L^2(-\infty, \infty)$ can be uniquely expressed as the basis of HPs in the following way

$$x(t) = \sum_{j=0}^{\infty} h_j H_j(t), \tag{15}$$

where h_j are the series coefficients that can be computed using (12) as

$$h_j = \frac{1}{\sqrt{\pi}2^j\Gamma(j+1)} \int_{-\infty}^{\infty} w(t)x(t)H_j(t)\mathrm{d}t, j = 0, 1, \cdots. \tag{16}$$

Considering the first $m+1$-terms of HPs, (15) can also be written as

$$x(t) \simeq \sum_{j=0}^{m} h_j H_j(t) = \chi^T\Omega(t), \tag{17}$$

where $\chi^T = [h_0, h_1, \cdots, h_m]$ and $\Omega(t) = [H_0(t), H_1(t), \cdots, H_m(t)]^T$.

4. Operational matrices of HPs

This section deals with the operational matrices of HPs that are constructed by using the analytical form (10) of HPs and Riemann–Liouville fractional-order integral operator and Hilfer fractional-order derivative operator.

Lemma 4.1. ([56], Section 3) *The Hermite integer-order integral operational matrix can be determined by using the following integral property*

$$\underbrace{\int_0^t \int_0^t \cdots \int_0^t}_{k-times} \Omega(y)(\mathrm{d}y)^k \simeq \mathbf{P}^k \Omega(t), \tag{18}$$

where $\mathbf{P}$ is the $m+1 \times m+1$ Hermite operational matrix of integration. For example, for $m = 4$, and $k = 1$, we have

$$\mathbf{P} = \begin{pmatrix} 0 & \frac{1}{2} & 0 & 0 & 0 \\ \frac{1}{2} & 0 & \frac{1}{4} & 0 & 0 \\ 0 & 0 & 0 & \frac{1}{6} & 0 \\ -\frac{3}{2} & 0 & 0 & 0 & \frac{1}{8} \\ 0 & 0 & 0 & 0 & 0 \end{pmatrix}, \tag{19}$$

$$\Omega(t) = \begin{pmatrix} 1 \\ 2t \\ 4t^2 - 2 \\ 8t^3 - 12t \\ 16t^4 - 48t^2 + 12 \end{pmatrix}, \tag{20}$$

and

$$\chi^T = \begin{pmatrix} \frac{1}{2} & 0 & \frac{1}{4} & 0 & 0 \end{pmatrix}. \tag{21}$$

Using (19)–(21), the approximate integral of t^2 is $\frac{t^3}{3}$ that coincides with the analytical integral of t^2.

Lemma 4.2. ([57], Thm. 1) *The Hermite integer-order derivative operational matrix can be determined by using the following derivative property*

$$\frac{d^k}{dt^k}\Omega(t) = \left(\mathbf{H}^{(1)}\right)^k \Omega(t), \tag{22}$$

where $\mathbf{H}^{(1)}$ is the $m+1 \times m+1$ Hermite operational matrix of derivatives. Defined as.

$$\mathbf{H}^{(1)}_{p,l} = \begin{cases} 2p, & for\, l = p-1, \\ 0, & \text{otherwise}. \end{cases}$$

For example at $m = 5$, we have

$$\mathbf{H}^{(1)} = \begin{pmatrix} 0 & 0 & 0 & 0 & 0 & 0 \\ 2 & 0 & 0 & 0 & 0 & 0 \\ 0 & 4 & 0 & 0 & 0 & 0 \\ 0 & 0 & 6 & 0 & 0 & 0 \\ 0 & 0 & 0 & 8 & 0 & 0 \\ 0 & 0 & 0 & 0 & 10 & 0 \end{pmatrix}, \tag{23}$$

$$\Omega(t) = \begin{pmatrix} 1 \\ 2t \\ 4t^2 - 2 \\ 8t^3 - 12t \\ 16t^4 - 48t^2 + 12 \\ 32t^5 - 160t^3 + 120t \end{pmatrix}, \tag{24}$$

and

$$\chi^T = (\,0.7788 \quad -0.0000 \quad -0.0974 \quad 0.0000 \quad 0.0020 \quad 0.0000\,). \tag{25}$$

Using (23)-(25), the approximate derivative of $\cos(t)$ is as following

$$\frac{d}{dt}\cos(t) \simeq 4t^4 + 0.12t^3 - 0.000000000000000022t^2 - 0.97t - 0.000000000000000010.$$

Lemma 4.3. *For $\gamma \in \mathbb{R}_+$ and $r, j \in \mathbb{N}$, we have the following result*

$$t^{j-2r+\gamma} \simeq \sum_{l=0}^{m} d_l H_l(t),$$

$$d_l = \frac{1}{2^l l! \sqrt{\pi}} \sum_{s=0}^{\lfloor \frac{l}{2} \rfloor} \frac{(-1)^s l! 2^{l-2s}}{s!(l-2s)!} \times \Gamma\left(\frac{l - 2(s+r) + j + \gamma + 1}{2}\right).$$

Proof. Approximating $t^{j-2r+\gamma}$ by using the $m+1$-terms of HPs as

$$t^{j-2r+\gamma} \simeq \sum_{l=0}^{m} d_l H_l(t). \tag{26}$$

The series coefficients d_l can be computed by using (16) as

$$\begin{aligned} d_l &= \frac{1}{2^l l! \sqrt{\pi}} \int_{-\infty}^{\infty} t^{j-2r+\gamma} H_l(t) w(t) \mathrm{d}t \\ &= \frac{1}{2^l l! \sqrt{\pi}} \sum_{s=0}^{\lfloor \frac{l}{2} \rfloor} \frac{(-1)^s l! 2^{l-2s}}{s!(l-2s)!} \int_{-\infty}^{\infty} t^{l-2(s+r)+j+\gamma} \exp\left(-t^2\right) \mathrm{d}t. \end{aligned} \tag{27}$$

Using Lemma 2.3, we can write Eq. (27) as

$$d_l = \frac{1}{2^l l! \sqrt{\pi}} \sum_{s=0}^{\lfloor \frac{l}{2} \rfloor} \frac{(-1)^s l! 2^{l-2s}}{s!(l-2s)!} \times \Gamma\left(\frac{l - 2(s+r) + j + \gamma + 1}{2}\right). \tag{28}$$

Consequently, Eqs. (26) and (28) prove the result. □

Lemma 4.4. *For $\alpha \in \mathbb{R}_+$ and $r, j \in \mathbb{N}$, we have the following result*

$$t^{j-2r-\alpha} \simeq \sum_{l=0}^{m} f_l H_l(t),$$

$$f_l = \frac{1}{2^l l! \sqrt{\pi}} \sum_{s=0}^{\lfloor \frac{l}{2} \rfloor} \frac{(-1)^s l! 2^{l-2s}}{s!(l-2s)!} \times \Gamma\left(\frac{l-2(s+r)+j-\alpha+1}{2}\right).$$

Proof. Approximating $t^{j-2r-\alpha}$ by using the $m+1$-terms of HPs as

$$t^{j-2r-\alpha} \simeq \sum_{l=0}^{m} f_l H_l(t). \tag{29}$$

The series coefficients f_l can be computed by using (16) as

$$\begin{aligned} f_l &= \frac{1}{2^l l! \sqrt{\pi}} \int_{-\infty}^{\infty} t^{j-2r-\alpha} H_l(t) w(t) \mathrm{d}t \\ &= \frac{1}{2^l l! \sqrt{\pi}} \sum_{s=0}^{\lfloor \frac{l}{2} \rfloor} \frac{(-1)^s l! 2^{l-2s}}{s!(l-2s)!} \int_{-\infty}^{\infty} t^{l-2(s+r)+j-\alpha} \exp(-t^2) \mathrm{d}t. \end{aligned} \tag{30}$$

Using Lemma 2.3, we can write Eq. (30) as

$$f_l = \frac{1}{2^l l! \sqrt{\pi}} \sum_{s=0}^{\lfloor \frac{l}{2} \rfloor} \frac{(-1)^s l! 2^{l-2s}}{s!(l-2s)!} \times \Gamma\left(\frac{l-2(s+r)+j-\alpha+1}{2}\right). \tag{31}$$

Consequently, Eqs. (29) and (31) prove the result. □

4.1 New Hermite generalized operational matrices

In this section, we introduce new operational matrices of HPs that are used to approximate the derivative terms of the problem (1). The operational matrices are constructed in the senses of the Riemann–Liouville fractional-order integral operator and Hilfer fractional-order derivative operator.

Theorem 4.5. *If $\Omega(t)$ is the Hermite function vector as defined in* Eq. (17), *then*

$${}_{RL}J_{0^+}^{\gamma} \Omega(t) \simeq \mathbf{P}^{(\gamma)} \Omega(t), \tag{32}$$

where $\mathbf{P}^{\gamma}$ is the Hermite generalized integral operational matrix of order $\gamma \in \mathbb{R}_+$ and dimensions $m+1 \times m+1$ that can be computed using the following expression

$$\mathbf{P}^{(\gamma)} = \sum_{l=0}^{m} \left(\sum_{r=0}^{\lfloor \frac{j}{2} \rfloor} \Psi_{j,l,r} \right), \quad j = 0, 1, \cdots, m, l = 0, 1, \cdots, m, \tag{33}$$

where

$$\Psi_{(j,l,r)} = \sum_{s=0}^{\lfloor \frac{l}{2} \rfloor} \frac{(-1)^{r+s} j! l! 2^{j-2r+l-2s} \Gamma\left(\frac{l-2(s+r)+j+\gamma+1}{2}\right)}{r! s! \Gamma(\gamma+j-2r+1) 2^l l! \sqrt{\pi}(l-2s)!}. \tag{34}$$

Proof. Applying the Riemann–Liouville fractional-order integral operator of order $\gamma \in \mathbb{R}_+$ defined in (5) to Eq. (10) and using (7), it yields

$${}_{\mathrm{RL}}J_{0^+}^{\gamma} H_j(t) = \sum_{r=0}^{\lfloor \frac{j}{2} \rfloor} \frac{(-1)^r j! 2^{j-2r} \Gamma(j-2r+1)}{r!(j-2r)! \Gamma(j-2r+1+\gamma)} t^{j-2r+\gamma}. \tag{35}$$

The term $t^{j-2r+\gamma}$ can be approximated by using basis of HPs as

$$t^{j-2r+\gamma} \simeq \sum_{l=0}^{m} d_l H_l(t). \tag{36}$$

Using the result of Lemma 4.3, (36) can be written as

$$t^{j-2r+\gamma} = \sum_{l=0}^{m} \left(\frac{1}{2^l l! \sqrt{\pi}} \sum_{s=0}^{\lfloor \frac{l}{2} \rfloor} \frac{(-1)^s l! 2^{l-2s}}{s!(l-2s)!} \times \Gamma\left(\frac{l-2(s+r)+j+\gamma+1}{2}\right) \right) H_l(t). \tag{37}$$

Using Eq. (37) in Eq. (35), we have

$$\begin{aligned} {}_{\mathrm{RL}}J_{0^+}^{\gamma} H_j(t) \simeq & \sum_{r=0}^{\lfloor \frac{j}{2} \rfloor} \frac{(-1)^r j! 2^{j-2r} \Gamma(j-2r+1)}{r!(j-2r)! \Gamma(j-2r+1+\gamma)} \times \\ & \sum_{l=0}^{m} \left(\frac{1}{2^l l! \sqrt{\pi}} \sum_{s=0}^{\lfloor \frac{l}{2} \rfloor} \frac{(-1)^s l! 2^{l-2s}}{s!(l-2s)!} \Gamma\left(\frac{l-2(s+r)+j+\gamma+1}{2}\right) \right) H_l(t). \end{aligned} \tag{38}$$

After simplification, we can write (38) as

$$\begin{aligned} \mathrm{RL}J_{0^+}^{\gamma} H_j(t) & \simeq \sum_{l=0}^{m} \left(\sum_{r=0}^{\lfloor \frac{j}{2} \rfloor} \sum_{s=0}^{\lfloor \frac{l}{2} \rfloor} \frac{(-1)^{r+s} j! l! 2^{j-2r+l-2s} \Gamma\left(\frac{l-2(s+r)+j+\gamma+1}{2}\right)}{r! s! \Gamma(j-2r+\gamma+1) 2^l l! \sqrt{\pi}(l-2s)!} \right) H_l(t) \\ & = \sum_{l=0}^{n} \left(\sum_{r=0}^{\lfloor \frac{j}{2} \rfloor} \Psi_{j,l,r} \right) H_l(t), \ j,l = 0,1,\cdots,m, \end{aligned} \tag{39}$$

where $\Psi_{j,l,r}$ is given in (34). Now (39) in vector form can be written as

$$ {}_{\mathrm{RL}}J^{\gamma}_{0^+}H_j(t) \simeq \left[\sum_{r=0}^{\lfloor \frac{j}{2} \rfloor} \Psi_{j,0,r}, \sum_{r=0}^{\lfloor \frac{j}{2} \rfloor} \Psi_{j,1,r}, \sum_{r=0}^{\lfloor \frac{j}{2} \rfloor} \Psi_{j,2,s} \cdots, \sum_{r=0}^{\lfloor \frac{j}{2} \rfloor} \Psi_{j,m,r} \right] \Omega(t). \tag{40} $$

Consequently, the required result is proved. □

For example, at $j, l = 0, 1, \cdots, 3$ and $\gamma = 3.5$, we have

$$ \mathbf{P}^{(3.5)} = \begin{pmatrix} 0.0275 & 0.0390 & 0.0240 & 0.0081 \\ 0.0173 & 0.0275 & 0.0195 & 0.0080 \\ -0.0350 & -0.0433 & -0.0206 & -0.0033 \\ -0.0720 & -0.1049 & -0.0650 & -0.0206 \end{pmatrix}. $$

Theorem 4.6. *If* $\Omega(t)$ *is the Hermite function vector as defined in* Eq. (17), *then*

$$ {}_{\mathrm{H}}\mathcal{D}^{\alpha,\beta}_{0^+}\Omega(t) \simeq \mathbf{H}^{(\alpha,\beta)}\Omega(t), \tag{41} $$

where $\mathbf{H}^{\alpha,\beta}$ is the Hermite generalized derivative operational matrix of order α and dimensions $m+1 \times m+1$ that can be computed using the following expression

$$ \mathbf{H}^{(\alpha,\beta)} = \sum_{l=0}^{m} \left(\sum_{r=0}^{\lfloor \frac{j-\lceil \alpha \rceil}{2} \rfloor} \Phi_{j,l,r} \right), \quad j = \lceil \alpha \rceil, \cdots, m, l = 0, 1, \cdots, m, \tag{42} $$

where

$$ \Phi_{(j,l,r)} = \sum_{s=0}^{\lfloor \frac{l}{2} \rfloor} \frac{(-1)^{r+s} j! l! 2^{j-2r+l-2s} \Gamma\left(\frac{l-2(s+r)+j-\alpha+1}{2} \right)}{r! s! \Gamma(j-2r+1-\alpha) 2^l l! \sqrt{\pi}(l-2s)!}. \tag{43} $$

Proof. Applying the Hilfer fractional-order derivative operator of order α defined in Definition (1) to Eq. (10) and using Lemma 2.2, it yields

$$ {}_{\mathrm{H}}\mathcal{D}^{\alpha,\beta}_{0^+}H_j(t) = \sum_{r=0}^{\lfloor \frac{j-\lceil \alpha \rceil}{2} \rfloor} \frac{(-1)^r j! 2^{j-2r} \Gamma(j-2r+1)}{r!(j-2r)! \Gamma(j-2r+1-\alpha)} t^{j-2r-\alpha}, j = \lceil \alpha \rceil, \lceil \alpha \rceil + 1, \cdots, m. \tag{44} $$

The term $t^{j-2r-\alpha}$ can be approximated by using basis of HPs as

$$ t^{j-2r-\alpha} \simeq \sum_{l=0}^{m} f_l H_l(t). \tag{45} $$

Using the result of Lemma 4.4, Eq. (45) can be written as

$$ t^{j-2r-\alpha} = \sum_{l=0}^{m} \left(\frac{1}{2^l l! \sqrt{\pi}} \sum_{s=0}^{\lfloor \frac{l}{2} \rfloor} \frac{(-1)^s l! 2^{l-2s}}{s!(l-2s)!} \times \Gamma\left(\frac{l-2(s+r)+j-\alpha+1}{2} \right) \right) H_l(t). \tag{46} $$

Using Eq. (46) in Eq. (44), we have

$$
{}_{\mathrm{H}}\mathcal{D}_{0^+}^{\alpha,\beta}H_j(t) \simeq \sum_{r=0}^{\lfloor\frac{j-\lceil\alpha\rceil}{2}\rfloor} \frac{(-1)^r j! 2^{j-2r}\Gamma(j-2r+1)}{r!(j-2r)!\Gamma(j-2r+1-\alpha)} \times \sum_{l=0}^{m}\left(\frac{1}{2^l l!\sqrt{\pi}}\sum_{s=0}^{\lfloor\frac{l}{2}\rfloor}\frac{(-1)^s l! 2^{l-2s}}{s!(l-2s)!}\Gamma\left(\frac{l-2(s+r)+j-\alpha+1}{2}\right)\right)H_l(t). \tag{47}
$$

After simplification, we can write (47) as

$$
\begin{aligned}
\mathrm{H}\mathcal{D}_{0^+}^{\alpha,\beta}H_j(t) &\simeq \sum_{l=0}^{m}\left(\sum_{r=0}^{\lfloor\frac{j-\lceil\alpha\rceil}{2}\rfloor}\sum_{s=0}^{\lfloor\frac{l}{2}\rfloor}\frac{(-1)^{r+s}j!l!2^{j-2r+l-2s}\Gamma\left(\frac{l-2(s+r)+j-\alpha+1}{2}\right)}{r!s!\Gamma(j-2r-\alpha+1)2^l l!\sqrt{\pi}(l-2s)!}\right)H_l(t) \\
&= \sum_{l=0}^{m}\left(\sum_{r=0}^{\lfloor\frac{j-\lceil\alpha\rceil}{2}\rfloor}\Phi_{j,l,r}\right)H_l(t), \; j=\lceil\alpha\rceil,\cdots,m, l=0,1,\cdots,m,
\end{aligned} \tag{48}
$$

where $\Phi_{j,l,r}$ is given in (43). Now (48) in vector form can be written as

$$
{}_{\mathrm{H}}\mathcal{D}_{0^+}^{\alpha,\beta}H_j(t) \simeq \left[\sum_{r=0}^{\lfloor\frac{j-\lceil\alpha\rceil}{2}\rfloor}\Phi_{j,0,r}, \sum_{r=0}^{\lfloor\frac{j-\lceil\alpha\rceil}{2}\rfloor}\Phi_{j,1,r}, \sum_{r=0}^{\lfloor\frac{j-\lceil\alpha\rceil}{2}\rfloor}\Phi_{j,2,s}\cdots, \sum_{r=0}^{\lfloor\frac{j-\lceil\alpha\rceil}{2}\rfloor}\Phi_{j,m,r}\right]\Omega(t). \tag{49}
$$

Consequently, the required result is proved. □

For example, at $j, l = 0, 1, \cdots, 3$, $\alpha = 1.5$, and $\beta = 1$, we have

$$
\mathbf{H}^{(1.5,1)} = \begin{pmatrix} 0 & 0 & 0 & 0 \\ 0 & 0 & 0 & 0 \\ 3.1205 & 2.3081 & 0.3901 & -0.0962 \\ 9.2325 & 9.3615 & 3.4622 & 0.3901 \end{pmatrix}.
$$

5. Applications of Hermite operational matrices

In this section, we develop a numerical algorithm that is based on the Hermite integrals and derivatives operational matrices. The framework of the proposed algorithm transforms the problem (1) to matrix equations of Sylvester types that are easy to handle with any computational software. The matrix equations compute the unknown vector χ^T which leads to the solution of the problem (1).

Suppose the following holds true

$$_{\mathrm{H}}\mathcal{D}^{\alpha,\beta}x(t) = \chi^T\Omega(t). \tag{50}$$

Integrating the Eq. (50) by applying the Riemann–Liouville fractional-order integral defined in (5) of order γ, we have

$$x(t) = \chi^T_{\mathrm{RL}}J^{\gamma}\Omega(t) + \sum_{a=0}^{1} b_a t^a, \quad 1 < \gamma \leq 2, 0 < \beta \leq 1, \tag{51}$$

where b_a's are the constant of integration determined by using the initial conditions (1), we have the following equation

$$x(t) = \chi^T_{\mathrm{RL}}J^{\gamma}\Omega(t) + \sum_{a=0}^{1} c_a t^a. \tag{52}$$

Using Theorem 4.5 and approximating the term $\sum_{a=0}^{1} c_a t^a$ with Hermite function vector, Eq. (51) can also be expressed as

$$x(t) \simeq \chi^T\mathbf{P}^{(\gamma)}\Omega(t) + B^T_{(1\times m+1)}\Omega(t), \tag{53}$$

where $\sum_{a=0}^{1} c_a t^a = B^T_{(1\times m+1)}\Omega(t)$. The terms of the problem (1) can be computed by using Theorem 4.6 and Eq. (53), we have

$$\begin{cases} _{\mathrm{H}}\mathcal{D}^{\alpha,\beta}x(t) = \chi^T\mathbf{P}^{(\gamma)}\mathbf{H}^{(\alpha,\beta)}\Omega(t) + B^T_{(1\times m+1)}\mathbf{H}^{(\alpha,\beta)}\Omega(t), \\ y(t) = A^T_{(1\times m+1)}\Omega(t). \end{cases} \tag{54}$$

Using Eqs. (50), (53), and (54) in (1), we have the following matrix equation of Sylvester type with an unknown vector χ^T of dimensions $1 \times m + 1$.

$$\chi^T\Omega(t) + \chi^T\left(\lambda_1\mathbf{P}^{(\gamma)}\mathbf{H}^{(\alpha,\beta)} + \lambda_2\mathbf{P}^{(\gamma)}\right)\Omega(t) = \left(A^T_{(1\times m+1)} - \lambda_1 B^T_{(1\times m+1)}\mathbf{H}^{(\alpha,\beta)} - \lambda_2 B^T_{(1\times m+1)}\right)\Omega(t) \tag{55}$$

By introducing the notations, $\Delta_{(m+1\times m+1)} = \lambda_1\mathbf{P}^{(\gamma)}\mathbf{H}^{(\alpha,\beta)} + \lambda_2\mathbf{P}^{(\gamma)}$ and $\Lambda_{(1\times m+1)} = A^T_{(1\times m+1)} - \lambda_1 B^T_{(1\times m+1)}\mathbf{H}^{(\alpha,\beta)} - \lambda_2 B^T_{(1\times m+1)}$ for the sake of simplifications, Eq. (55) can be written as

$$\chi^T_{(1\times m+1)} + \chi^T_{(1\times m+1)}\Delta_{(m+1\times n+1)} = \Lambda_{(1\times m+1)}. \tag{56}$$

By solving (56), we can easily compute the unknown vector $\chi^T_{(1\times m+1)}$ which then substituting in Eq. (53) yields the solution of the problem (1).

6. Examples

In this section, we solve some examples to test the applicability and efficiency of the proposed algorithm discussed in Section 5. The results are displayed in Tables and Plots.

Example 6.1. Consider the following Bagley–Torvik equation with initial conditions

$$\begin{aligned} &\lambda_3 x''(t) + \lambda_{1\mathrm{H}}\mathcal{D}^{\alpha,\beta}x(t) + \lambda_2 x(t) = y(t), t \in (-10, 10), \\ &x(0) = c_1, \quad x'(0) = c_2, \end{aligned} \tag{57}$$

where, $y(t) = 1 + t$, $1 < \alpha < 2$, $0 < \beta < 1$, $c_1 = 0$, $c_2 = 1$, and $\lambda_1 = \lambda_2 = \lambda_3 = 1$. At $m = 2$, we

$$\mathbf{P}^{(2)} = \begin{pmatrix} \frac{1}{8} & \frac{5081767996463981}{36028797018963968} & \frac{1}{16} \\ \frac{6775690661951975}{72057594037927936} & \frac{1}{8} & \frac{5081767996463981}{72057594037927936} \\ -\frac{1}{8} & -\frac{6775690661951975}{72057594037927936} & 0 \end{pmatrix}, \tag{58}$$

$$\mathbf{H}^{(1.5,1)} = \begin{pmatrix} 0 & 0 & 0 \\ 0 & 0 & 0 \\ \frac{7026736834630197}{2251799813685248} & \frac{5197457860426675}{2251799813685248} & \frac{7026736834630197}{18014398509481984} \end{pmatrix}, \tag{59}$$

$$\chi^T = \left(-\frac{1917597440026193}{2596148429267413814265248164610048} \quad -\frac{5162427874821681}{10384593717069655257060992658440192} \quad -\frac{734167200735167}{811296384146066816957890051440064} \right), \tag{60}$$

$$\Omega(t) = \begin{pmatrix} 1 \\ 2t \\ 4t^2 - 2 \end{pmatrix}. \tag{61}$$

Using Eqs. (58)–(61), the approximated solution, $\hat{x(t)}$ of the Example 6.1 is $1 + t$, which coincides with its exact solution, $x(t) = 1 + t$ at $\alpha = \frac{3}{2}$ and $\beta = 1$.

Example 6.2. Consider the following Bagley–Torvik equation with initial conditions

$$\begin{aligned} &\lambda_3 x''(t) + \lambda_{1e\mathrm{H}}\mathcal{D}^{\alpha,\beta}x(t) + \lambda_2 x(t) = y(t), t \in (-2, 2), \\ &x(0) = c_1, \quad x'(0) = c_2, \end{aligned} \tag{62}$$

where, $y(t) = \lambda_2(1 + t)$, $1 < \alpha < 2$, $0 < \beta < 1$, $c_1 = 0$, $c_2 = 1$, $\lambda_1 = 1.5$, $\lambda_2 = 2.5$, and $\lambda_3 = 1$.

Example 6.3. Consider the following Bagley–Torvik equation with initial conditions

$$\begin{aligned} &\lambda_3 x''(t) + \lambda_{1\mathrm{H}}\mathcal{D}^{\alpha,\beta}x(t) + \lambda_2 x(t) = y(t), t \in (0, 1), \\ &x(0) = c_1, \quad x'(0) = c_2, \end{aligned} \tag{63}$$

where, $y(t) = 8$, $1 < \alpha < 2$, $0 < \beta < 1$, $c_1 = 0 = c_2$, $\lambda_3 = 1$, and $\lambda_1 = 0.5 = \lambda_2$. The analytical solution of Example 6.3 is given as

$$x(t) = \int_0^t G(t - z)y(z)dz, \tag{64}$$

where G is a Green function given as under

$$G(t) = \sum_{l=0}^{\infty} \frac{(-1)^l}{\Gamma(l+1)} \lambda_2^l t^{2l+1} E^{(l)}_{1/2,2+3l/2}\left(-\lambda_1 \sqrt{t}\right), \tag{65}$$

The expression $E^{(l)}_{\gamma,\delta}$ is the lth derivative of the two parametric Mittagg-Leffler function, given as ([26], (8.26))

$$E^{(l)}_{\gamma,\delta}(t) = \sum_{i=0}^{\infty} \frac{\Gamma(i+l+1)t^i}{\Gamma(i+1)\Gamma(\gamma i + \gamma l + \delta)}, \quad l = 0, 1, \cdots. \tag{66}$$

Example 6.4. Consider the following Bagley–Torvik equation with initial conditions

$$\begin{aligned} &\lambda_3 x^{(\gamma)}(t) + \lambda_1 {}_{\mathrm{H}}\mathcal{D}^{\alpha,\beta} x(t) + \lambda_2 x(t) = y(t), t \in (0,2), \\ &x(0) = c_1, \ \ x'(0) = c_2 \end{aligned} \tag{67}$$

where, $y(t) = 0$, $0 < \gamma < 2$, $c_1 = 1$, $c_2 = 0$, $\lambda_3 = 1$, $\lambda_1 = 0$, and $\lambda_2 = 1$. The exact solution of Example 6.4 at $\gamma = 2$ is, $x(t) = \cos(t)$.

7. Discussion

We solved the Bagley–Torvik FDDE by developing the fractional-order derivative operational matrices of HPs. The operational matrices were developed in the sense of Hilfer fractional-order derivative. We observed that HPs' basis is well fit to approximate any square-integrable function on the entire real line, see **Figures 1** and **2**. Also, the integrals and derivatives of square-integrable functions were computed by using the Hermite operational matrices had a great resemblance with the results computed by using the analytical techniques of integrals and derivatives, see **Figures 3** and **4**. Based on the Hermite operational matrices, we introduced a numerical algorithm that is capable to transform the FDDE into a system of matrix equations of Sylvester types that are easy to handle with any computational software. We checked the accuracy

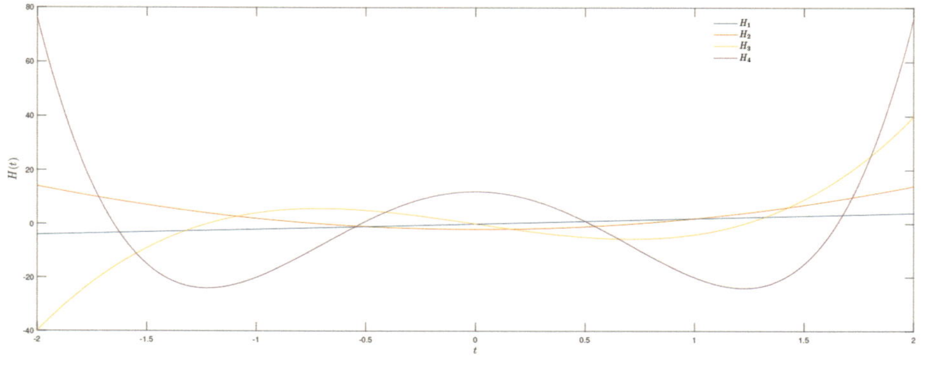

Figure 1.
Hermite polynomials plots for various j.

DOI: http://dx.doi.org/10.5772/intechopen.1001144

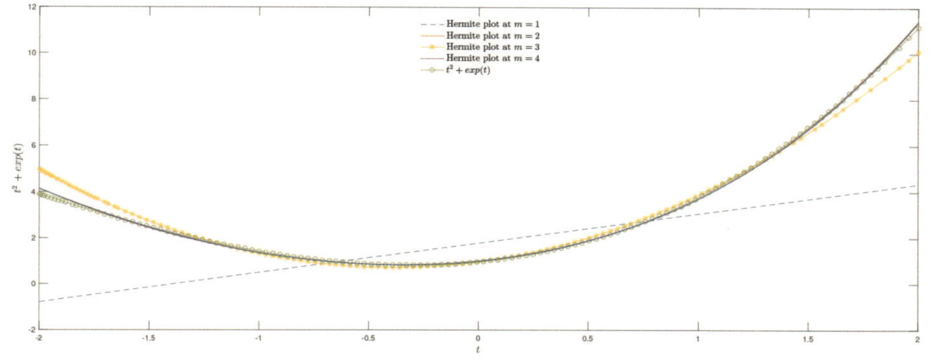

Figure 2.
Approximation of $t^2 + \exp(t)$ using Hermite function vectors (17) at various values of m.

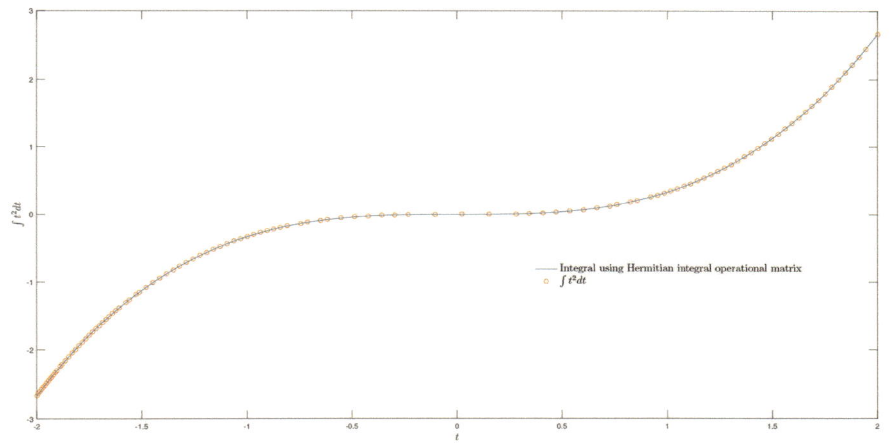

Figure 3.
The analytical and approximate integral plots of t^2 at $m = 4$ by using Hermite integral operational matrix.

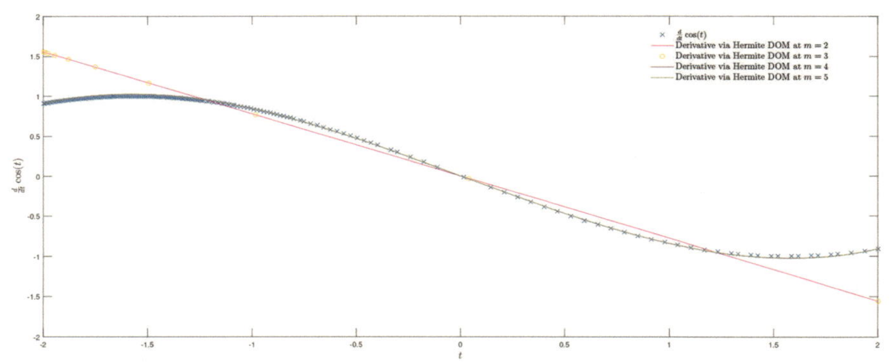

Figure 4.
The analytical and approximate derivative plots of $\cos(t)$ at different values of m by using Hermite derivative operational matrices (DOM).

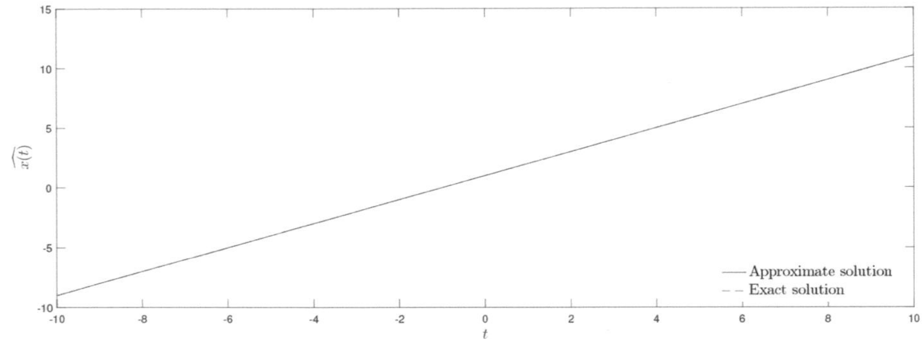

Figure 5.
The graphical view of exact and approximate solutions of Example 6.1 at $m = 2$, and $\alpha = 1.5$.

t	$\hat{x(t)}$ at $m = 2$	$\hat{x(t)}$ at $m = 3$	$\hat{x(t)}$ at $m = 10$	$\hat{x(t)}$ at $m = 15$	Exact solution
−1	0.0	0.0	0.0	0.0	0.0
−0.8	0.2	0.2	0.2	0.2	0.2
−0.6	0.4	0.4	0.4	0.4	0.4
−0.4	0.6	0.6	0.6	0.6	0.6
−0.2	0.8	0.8	0.8	0.8	0.8
0	1.0	1.0	1.0	1.0	1.0
0.2	1.2	1.2	1.2	1.2	1.2
0.4	1.4	1.4	1.4	1.4	1.4
0.6	1.6	1.6	1.6	1.6	1.6
0.8	1.8	1.8	1.8	1.8	1.8
1	2.0	2.0	2.0	2.0	2.0

Table 1.
Approximate solution of Example 6.2 is computed at various m.

and stability of the PNA by solving various Bagley–Torvic types FDDE corresponding to various initial conditions. We observed that the approximate solution obtained by using the PNA coincided with the exact solution by taking only a few terms of HPs, see (Example 6.1, **Figure 5**) and (Example 6.2, **Table 1**). We also analyzed the stability of the PNA by computing the approximate solution at various values of α and at various values of m, see **Figures 6–10**, and **Table 2**. We noted that as m getting large and α was getting closer to $\frac{3}{2}$, the approximate solution approached to the exact solution of the problem. We also computed the amount of the absolute error for Example 6.4 at various values of m, and observed that the error decreased significantly for increasing m, see **Table 3**. The numerical accuracy of the results computed by using PNA was also analyzed by comparing the results with the Adomian method. We noted that the PNA prodoced better accuracy as compared to the Adomian method, see (Example 6.3, **Table 2**).

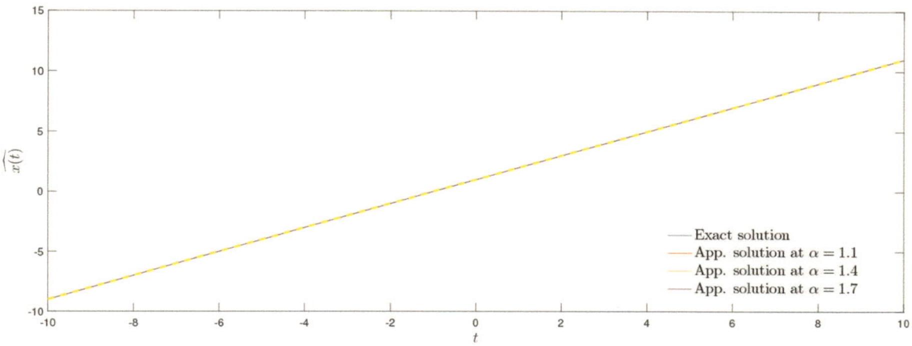

Figure 6.
The graphical view of exact and approximate solutions of Example 6.1 at $m = 2$, and various values of α.

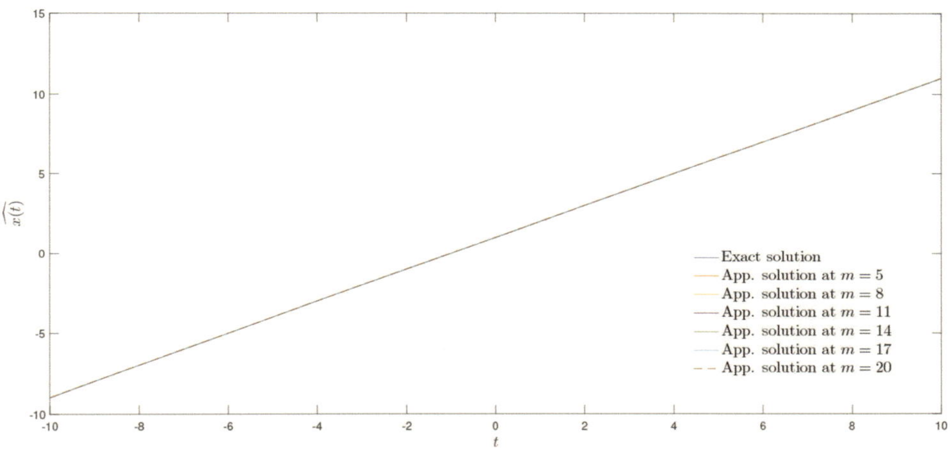

Figure 7.
The graphical view of exact and approximate solutions of Example 6.1 at $\alpha = 1.5$, and various values of m.

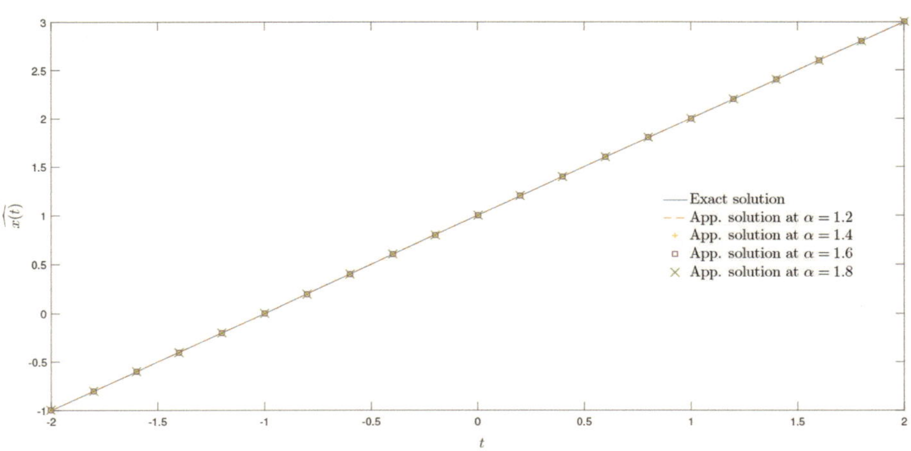

Figure 8.
The graphical view of exact and approximate solutions of Example 6.2 at $m = 2$, and various values of α.

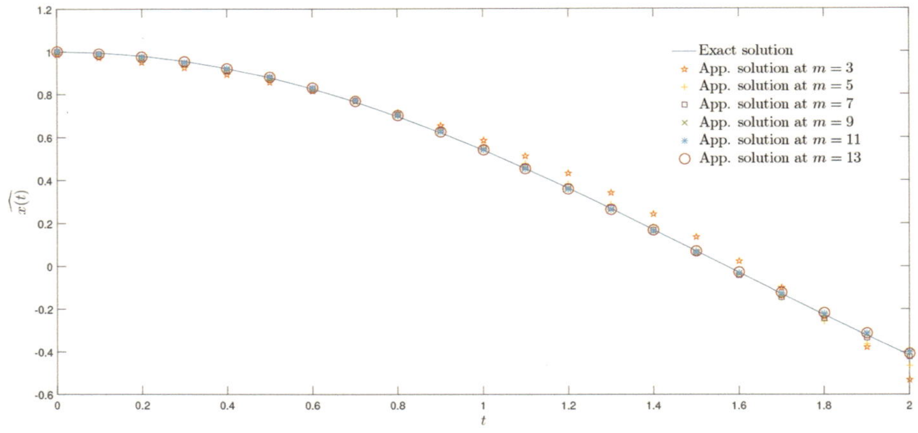

Figure 9.
The graphical view of exact and approximate solutions of Example 6.4 at various values of m.

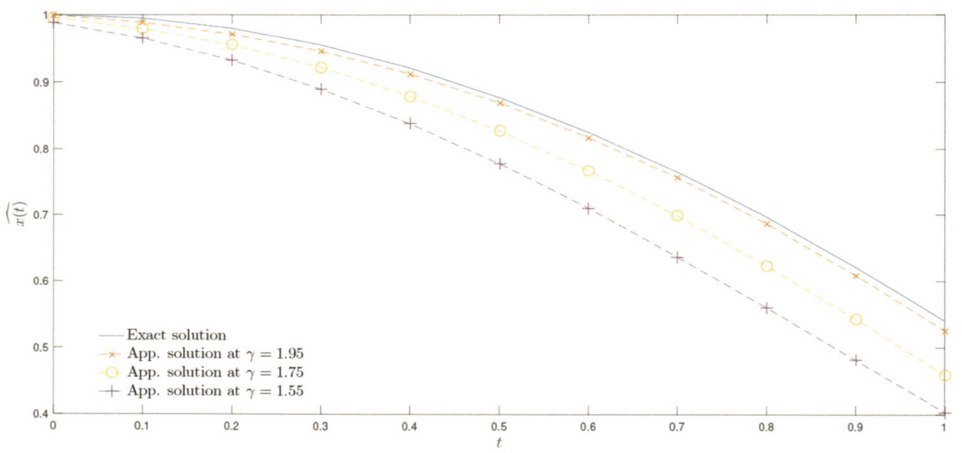

Figure 10.
The graphical view of exact and approximate solutions of Example 6.4 at $m = 12$, and various values of γ.

t	Analytical solution	Adomian method	PNA at $m = 16$	PNA at $m = 20$
0	0.000000	0.000000	0.000000	0.000000
0.2	0.125221	0.140640	0.168975	0.147779
0.4	0.455435	0.533284	0.509874	0.491041
0.6	0.950392	1.148840	1.0384000	1.024804
0.8	1.579557	1.963033	1.726857	1.693681
1	2.315526	2.952567	2.524754	2.453043

Table 2.
Approximate solution of Example 6.3 obtained using the proposed numerical algorithm (PNA) are compared with the solution obtained using the Adomian method [32].

t	Error at $m = 5$	Error at $m = 7$	Error at $m = 9$	Error at $m = 15$	Error at $m = 30$
0	0.000851555699954	0.000015168258275	0.000038589837968	0.000011170515838	0.0000000771779988
0.2	0.013708145743312	0.009294768239541	0.007147462915103	0.003988632129512	0.0000055034808212
0.4	0.011893816308316	0.006240104002812	0.003411075508628	0.000010952382595	0.0000000030115944
0.6	0.001779738199402	0.001752762045013	0.003127688012188	0.003162961831284	0.0000009796683838
0.8	0.010343264247858	0.008577085429824	0.006809926671596	0.002261792740019	0.0000000600336283
1	0.018882601807647	0.010311814708410	0.005334232561979	0.000946037042348	0.0000000892574348

Table 3.
Absolute error of Example 6.4 is computed at various m.

Classification

2020 Mathematics Subject Classification. 65 L05, 26A33, 35R11

Author details

Imran Talib[1*] and Faruk Özger[2]

1 Nonlinear Analysis Group (NAG), Mathematics Department, Virtual University of Pakistan, Pakistan

2 Department of Engineering Sciences, Izmir Katip Celebi University, Izmir, Turkey

*Address all correspondence to: imrantaalib@gmail.com; imrantalib@vu.edu.pk

References

[1] Conway JB. A Course in Functional Analysis. Vol. 96. New York, NY: Springer; 2019

[2] Campbell PJ. The origin of "zorn's lemma". Historia Mathematica. 1978; **5**(1):77-89

[3] Blass A. Existence of bases implies the axiom of choice. Contemporary Mathematics. 1984;**31**:1-3. DOI: 10.1090/conm/031/763890

[4] Keremedis K. Bases for vector spaces over the two-element field and the axiom of choice. Proceedings of the American Mathematical Society. 1996; **124**(8):2527-2531

[5] Talib I, Noor ZA, Hammouch Z, Khalil H. Compatibility of the paraskevopoulos's algorithm with operational matrices of vieta–lucas polynomials and applications. Mathematics and Computers in Simulation. 2022;**202**:442-463

[6] Talib I, Raza A, Atangana A, Riaz MB. Numerical study of multi-order fractional differential equations with constant and variable coefficients. Journal of Taibah University for Science. 2022;**16**(1):608-620

[7] Talib I, Bohner M. Numerical study of generalized modified caputo fractional differential equations. International Journal of Computer Mathematics. 2022; **100**:1-24

[8] Kumar N, Mehra M. Collocation method for solving nonlinear fractional optimal control problems by using hermite scaling function with error estimates. Optimal Control Applications and Methods. 2021;**42**(2):417-444

[9] Kumar S, Kumar R, Momani S, Hadid S. A study on fractional covid-19 disease model by using hermite wavelets. Mathematical Methods in the Applied Sciences. 2021:1-17. DOI: 10.1002/mma.7065

[10] Yari A. Numerical solution for fractional optimal control problems by hermite polynomials. Journal of Vibration and Control. 2021;**27**(5–6): 698-716

[11] Talib I, Jarad F, Mirza MU, Nawaz A, Riaz MB. A generalized operational matrix of mixed partial derivative terms with applications to multi-order fractional partial differential equations. Alexandria Engineering Journal. 2022; **61**(1):135-145

[12] Pu T, Fasondini M. The numerical solution of fractional integral equations via orthogonal polynomials in fractional powers. 2022. arXiv preprint arXiv: 2206.14280

[13] Doha EH, Bhrawy AH, Ezz-Eldien SS. A new Jacobi operational matrix: An application for solving fractional differential equations. Applied Mathematical Modelling. 2012;**36**(10): 4931-4943

[14] Doha EH, Bhrawy AH, Ezz-Eldien SS. Efficient Chebyshev spectral methods for solving multi-term fractional orders differential equations. Applied Mathematical Modelling. 2011; **35**(12):5662-5672

[15] Kazem S, Abbasbandy S, Kumar S. Fractional-order Legendre functions for solving fractional-order differential equations. Applied Mathematical Modelling. 2013;**37**(7):5498-5510

[16] Talaei Y, Asgari M. An operational matrix based on Chelyshkov polynomials for solving multi-order fractional

differential equations numerical solution the fractional Bagley–Torvik equation arising in fluid mechanics. Neural Computing and Applications. 2018;**30**: 1369-1376

[17] Harris FE. Chapter 14 - series solutions: Important odes. In: Harris FE, editor. Mathematics for Physical Science and Engineering. Boston: Academic Press; 2014. pp. 487-543

[18] Van Assche W. Ordinary special functions. In: Françoise J-P, Naber GL, Tsun TS, editors. Encyclopedia of Mathematical Physics. Oxford: Academic Press; 2006. pp. 637-645

[19] Poteryaeva VA, Bubenchikov MA. Applications of orthogonal polynomials to solving the schrodinger equation. Reports on Mathematical Physics. 2022; **89**(3):307-317

[20] Dattoli G. Laguerre and generalized hermite polynomials: the point of view of the operational method. Integral Transforms and Special Functions. 2004; **15**(2):93-99

[21] Secer A, Ozdemir N, Bayram M. A hermite polynomial approach for solving the SIR model of epidemics. Mathematics. 2018;**6**(12):305

[22] Perote J, Del Brio E. Positive definiteness of multivariate densities based on hermite polynomials. Available at SSRN 672522. 2005

[23] Zhang X-Y, Zhao Y-G, Zhao-Hui L. Straightforward hermite polynomial model with application to marine structures. Marine Structures. 2019;**65**:362-375

[24] Liu M, Peng L, Huang G, Yang Q, Jiang Y. Simulation of stationary non-gaussian multivariate wind pressures using moment-based piecewise hermite polynomial model. Journal of Wind Engineering and Industrial Aerodynamics. 2020;**196**:104041

[25] Diethelm K, Ford J. Numerical solution of the Bagley-Torvik equation. BIT Numerical Mathematics. 2002; **42**(3):490-507

[26] Podlubny I. Fractional differential equations, volume 198 of Mathematics in Science and Engineering. In: An Introduction to Fractional Derivatives, Fractional Differential Equations, to Methods of their Solution and Some of their Applications. San Diego, CA: Academic Press, Inc.; 1999

[27] Esmaeili S. The numerical solution of the Bagley-Torvik equation by exponential integrators. Scientia Iranica. 2017;**24**(6):2941-2951

[28] Al-Mdallal QM, Syam MI, Anwar MN. A collocation-shooting method for solving fractional boundary value problems. Communications in Nonlinear Science and Numerical Simulation. 2010;**15**(12):3814-3822

[29] Gülsu M, Öztürk Y, Anapali A. Numerical solution the fractional Bagley–Torvik equation arising in fluid mechanics. International Journal of Computer Mathematics. 2017;**94**(1):173-184

[30] Yüzbaş Ş. Numerical solution of the Bagley–Torvik equation by the bessel collocation method. Mathematical Methods in the Applied Sciences. 2013; **36**(3):300-312

[31] Setia A, Liu Y, Vatsala AS. The solution of the Bagley-Torvik equation by using second kind chebyshev wavelet. In: 2014 11th International Conference on Information Technology: New Generations. Las Vegas: IEEE; 2014. pp. 443-446

[32] Momani S, Odibat Z. Numerical comparison of methods for solving linear

differential equations of fractional order. Chaos, Solitons & Fractals. 2007;**31**(5): 1248-1255

[33] Saha Ray S, Bera RK. Analytical solution of the bagley torvik equation by adomian decomposition method. Applied Mathematics and Computation. 2005;**168**(1):398-410

[34] Luchko Y, Gorenflo R. The Initial Value Problem for Some Fractional Differential Equations with the Caputo Derivatives. Berlin: Fachbereich Mathematik und Informatik, Freie Universität Berlin; 1998 Preprint Serie A 08-98

[35] Torvik PJ, Bagley RL. On the appearance of the fractional derivative in the behavior of real materials. Journal of Applied Mechanics. 1984;**51**:294-298

[36] Ji T, Hou J, Yang C. Numerical solution of the Bagley–Torvik equation using shifted chebyshev operational matrix. Adv. Difference Equ. 2020; **2020**(1):1-14

[37] Uddin M, Ahmad S. On the numerical solution of Bagley–Torvik equation via the laplace transform. Tbilisi Mathematical Journal. 2017;**10**(1): 279-284

[38] Rehman MU, Khan RA. A numerical method for solving boundary value problems for fractional differential equations. Applied Mathematical Modelling. 2012;**36**(3):894-907

[39] Saadatmandi A, Dehghan M. A new operational matrix for solving fractional-order differential equations. Computers & Mathematcs with Applications. 2010; **59**(3):1326-1336

[40] Oldham K, Spanier J. The Fractional Calculus Theory and Applications of Differentiation and Integration to Arbitrary Order. New York, NY: Elsevier; 1974

[41] Trujillo JJ, Scalas E, Diethelm K, Baleanu D. Fractional Calculus: Models and Numerical Methods. Vol. 5. London: World Scientific; 2016

[42] Ross B. A brief history and exposition of the fundamental theory of fractional calculus. Fractional Calculus and its Applications. 1975;**457**:1-36

[43] Tenreiro Machado J, Kiryakova V, Mainardi F. Recent history of fractional calculus. Communications in Nonlinear Science and Numerical Simulation. 2011; **16**(3):1140-1153

[44] Baleanu D, Agarwal RP. Fractional calculus in the sky. Adv. Difference Equ., pages Paper No. 117, 9. 2021

[45] Atangana A. Mathematical model of survival of fractional calculus, critics and their impact: How singular is our world? Adv. Difference Equ., pages Paper No. 403, 59. 2021

[46] Hilfer R. Applications of Fractional Calculus in Physics. River Edge, NJ: World Scientific Publishing Co., Inc.; 2000

[47] Atangana A, Baleanu D. New fractional derivative with non-local and non-singular kernel. Journal of Thermal Science. 2016;**20**(2):763-769

[48] Samko SG, Kilbas AA, Marichev OI. Fractional integrals and derivatives. In: Nikolskiĭ SM, editor. Translated from the 1987 Russian original, Revised by the authorsTheory and Applications. Edited and with a foreword by. Yverdon: Gordon and Breach Science Publishers; 1993

[49] Katugampola UN. New approach to a generalized fractional integral. Applied

Mathematics and Computation. 2011; **218**(3):860-865

[50] Katugampola UN. A new approach to generalized fractional derivatives. Bulletin of Mathematical Analysis and Applications. 2014;**6**(4):1-15

[51] Atangana A. Fractional Operators with Constant and Variable Order with Application to Geo-hydrology. New York, NY: Academic Press; 2017

[52] Goodwine B, Leyden K. Recent results in fractional-order modeling in multi-agent systems and linear friction welding. IFAC. 2015;**48**(1):380-381

[53] Goodwine B. Modeling a multi-robot system with fractional-order differential equations. In: 2014 IEEE International Conference on Robotics and Automation (ICRA). Hong Kong: IEEE; 2014. pp. 1763-1768

[54] Shloof AM, Senu N, Ahmadian A, Nik Long NMA, Salahshour S. Solving fractal-fractional differential equations using operational matrix of derivatives via hilfer fractal-fractional derivative sense. Applied Numerical Mathematics. 2022;**178**:386-403

[55] Bell WW. Special Functions for Scientists and Engineers. New Jersey: Van Nostrand; 1968

[56] Th Kekkeris G, Paraskevopoulos PN. Hermite series approach to optimal control. International Journal of Control. 1988;**47**(2):557-567

[57] Kalateh BZ, Ahmadi AS, Aminataei A. Operational matrices with respect to hermite polynomials and their applications in solving linear differential equations with variable coefficients. 2013

Chapter 4

Fitting Parametric Polynomials Based on Bézier Control Points

Mustafa Abbas Fadhel

Abstract

This chapter provides an overview at fitting parametric polynomials with control points coefficients. These polynomials have several properties, including flexibility and stability. Bézier, B-spline, Nurb, and Bézier trigonometric polynomials are the most significant of these kinds. These fitting polynomials are offered in two dimensions (2D) and three dimensions (3D). This type of polynomial is useful for enhancing mathematical methods and models in a variety of domains, the most significant of which being interpolation and approximation. The utilization of parametric polynomials minimizes the number of steps in the solution, particularly in programming, as well as the fact that polynomials are dependent on control points. This implies having more choices when dealing with the generated curves and surfaces in order to produce the most accurate results in terms of errors. Furthermore, in practical applications such as the manufacture of automobile exterior constructions and the design of surfaces in various types of buildings, this kind of polynomial has absolute preference.

Keywords: parametric polynomials, parametric curves, parametric surfaces, control points, parametric values

1. Introduction

Curve (or surface) fitting to a set of provided data points based on control points is a common and often performed activity in many areas of science and engineering, including machine vision, computer vision, reverse engineering, coordinate metrology, and computer-aided geometric design.

When fitting a collection of data points in two- and three-dimensional space, the shortest distance (which is also known as Euclidean distance, geometric distance, or orthogonal distance) between the curve (or surface) and the data points point has practical significance [1–3]. The fitting of a parametric curve or surface using this error measure is known as "orthogonal distance fitting" or "geometric fitting" [4–11]. The geometric fitting issue is a nonlinear minimization problem that must be addressed iteratively, and it is widely acknowledged as an analytically and computationally challenging problem. There are two new methods for geometric fitting of functional curves and surfaces: [4, 6, 9, 11] for implicit curves and surfaces and [10, 12, 13] for parametric curves and surfaces. For a long time, the theoretical and computational problems in computing and decreasing geometric distances challenged the development of geometric fitting algorithms for functional curves/surfaces [6, 7].

Most data processing software programs use approximation measures of geometric distance to avoid the challenges associated with geometric fitting [8, 14–18]. Nonetheless, when highly accurate and reliable estimation of parameter estimation is required by applications, we have no choice but to use geometric distance as the error measure for functions based on control points fitting, despite the high computational cost and difficulties in developing geometric fitting algorithms. In fact, an international standard requires the use of the geometric error measure for testing data processing software for coordinate metrology [2].

Ameer et al. [19] developed a new Bézier polynomial with two form parameters based on new generalized Bernstein basis functions. Both Bernstein basis functions and Bézier polynomials have geometric features that are analogous to the classical Bernstein basis and Bézier curve, respectively. Using the described Bézier curves and surfaces as applications, several free-form curves may be represented.

Khan et al. [20] investigate weighted Lupaş post-quantum Bernstein blending functions and Bézier curves built using bases through (p, q) integers. Because of their degree elevation qualities and the de Casteljau technique, quadratic weighted Lupaş post-quantum Bézier curves may represent conic sections in the two-dimensional plane. Compared to traditional rational Bézier curves, Lupaş q-Bézier curves, and weighted Lupaş q-Bézier curves, its generalization provides superior approximation and adaptability to a specific control point as well as a control polygon.

Özger [21] defined a new type of Bézier base using λ shape parameters. He creates a new type of Schurer operator by defining new Bézier-Schurer bases.

Fitting parametric Bézier-like curves and surfaces in two- and three-dimensional space based on control points is the focus of this chapter.

2. Parametric polynomials

In computer graphics, we frequently need to draw many types of objects onto the screen. Objects are not always flat, and we must draw curves several times in order to depict an object. A curve is a collection of infinitely many points. Except for endpoints, each point has two neighbors. Curves are broadly classified into three types: explicit, implicit, and parametric. A curve can be created for the mathematical function $y = f(x)$, which gives an explicit representation of the curve. The explicit representation is not general since it cannot express vertical lines and is also single-valued. Normally, the function computes only one value of y for each value of x.

The most basic practice for writing a curve equation is implicit representation, which combines all variables into one lengthy, non-linear equation, like as follows:

$$ax^3 + by^2 + 2cxy + 2dx + 2ey + f = 0. \tag{1}$$

To compute the values and plot them on a graph in this representation, we must solve the complete non-linear equation.

Some interactions between two numbers or variables are so complex that we occasionally incorporate a third quantity or variable to simplify matters. This third number is known as a parameter. Instead of a single equation relating, say, x and y, we have two, one related to the x parameter and one relating to the y parameter.

The parametric representation rewrites the equation into shorter, more readily solvable equations that transform the values of one variable into the values of the others:

$$x = a + bt + ct^2 + dt^3 \tag{2}$$

$$y = g + ht + jt^2 + kt^3 \tag{3}$$

The equations for x and y are straightforward when using this form. We just require a value for t, which is the point along the curve where we wish to compute x and y. Parametric curves can be imagined as being drawn by a point passing through space. We can determine the x and y values of the moving point at any value of t. This is a critical issue since many tools depend on the notion of associating a parameter number with each point on the line. This relates to the curve's U dimension.

3. Bézier curves

An explicit Bézier curves is one for which the x-coordinates of the control points are evenly spaced between 0 and 1. A Bézier curve takes on the important special form

$$x = t \tag{4}$$

$$y = f(t) \tag{5}$$

or simply

$$y = f(x). \tag{6}$$

An explicit Bézier curves is sometimes called a non-parametric Bézier curve and it developed by A French engineer named Pierre Bézier to define the computer graphics curve. This function generates a Bézier curve, the form of which is specified by control points. The inner control points (points not on the curve) determine the form of the curve, whereas the end control points specify the endpoints of the curve.

An nth-degree of Bezier curve is depicted by the formula below.

$$P(t) = \sum_{a=0}^{\theta} p_a \, B_a^{\theta}(t), \tag{7}$$

where $t \in [0, 1]$ is the parameter, n is the degree of Bezier curve, and $p_a = \left(p_x^{(a)}, p_y^{(a)}\right)$ are the control points.

The basis functions, are the parametric Bernstein polynomial of degree θ which are defined explicitly by

$$B_a^{\theta}(t) = \binom{\theta}{a} t^a (1 - t)^{\theta - a} \tag{8}$$

with $\binom{\theta}{a} = \frac{\theta!}{a!(\theta - a)!}$ and $\theta \in Z^+$. **Figure 1** shows a degree five explicit Bézier curve. Eq. (7) is alternatively parametrically expressed as

$$P(t) = (x(t), y(t)) \tag{9}$$

with

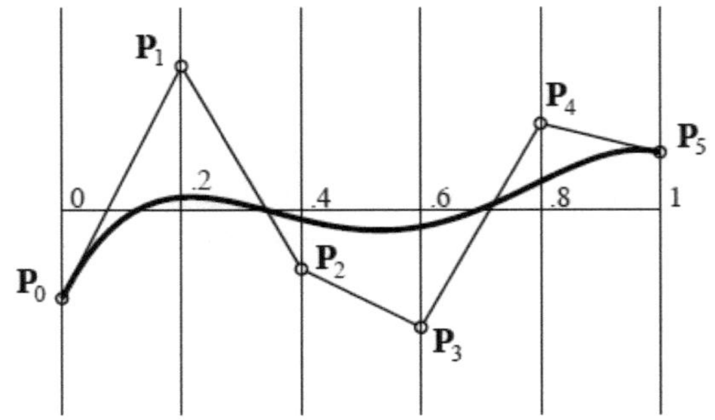

Figure 1.
Explicit Bézier curve.

$$x(t) = \sum_{a=0}^{\theta} p_x^{(a)} B_a^{\theta}(t) \tag{10}$$

and

$$y(t) = \sum_{a=0}^{\theta} p_y^{(a)} B_a^{\theta}(t). \tag{11}$$

Several construction-based Bezier control points are used in this work to find interpolation polynomials.

4. Fitting Bézier polynomials

Approximation and interpolation approaches can be used to fit the curve (or surface). However, classical interpolation and interpolation with control points are the most extensively utilized approaches [22]. Control points, with the exception of the end control points, are points that do not appear on the curves, but they regulate the form of the curves (or surfaces). In other words, control points function like magnets.

4.1. Fast Bézier interpolator

Yau and Wang [23] suggested a Fast Bézier interpolator approach for generating control points to produce a local Bézier interpolating polynomial. In this approach, the data points were categorized according to their distribution. The values of two inner control points might be obtained by meeting the stated requirements for each set of data points. The arc was formed by intersecting line segments between data points and applying the angle criteria. The inaccuracy from matching the data points with the arc was then used to determine the curve in each segment. Nevertheless, the mathematical procedure for determining the value of two inner control points for each subinterval is time-consuming due to the numerous computations required.

Figure 2 is a chart of cubic Bezier curve-fitted continuous short blocks (CSBs), often known as G_{01} blocks. $P_1 - P_8$ and $P_9 - P_n$ are curves made up of CSBs fitted by two cubic Bezier curves, respectively. The corner angles are y_1 and y_2. The lengths of the cubic Bezier curves are L_1 and L_3. L_2 denotes the length of a G_{01}

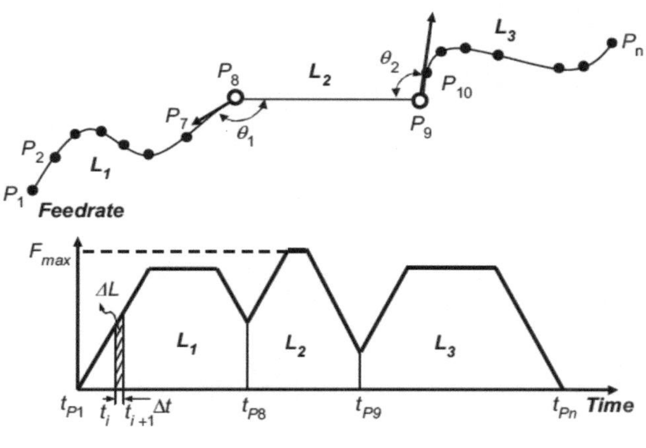

Figure 2.
CSBs with Cubic Bezier curves and a feedrate profile.

This figure describes how a linear segment set may be divided into several CSB areas and linear segment regions by break points P_8 and P_9. A corner error arising from system dynamics and geometry limits the corner feedrate. In comparison, the maximum feedrate for a cubic Bezier curve is calculated using the curvature, the permissible maximum machining speed, and the feedrate assigned. The trapezoidal feed rate profile is used to design feedrate acceleration and deceleration. The feed length per sample period, ΔL, is calculated from the resultant feedrate profile, and the interpolated locations are calculated.

The first step in using the CSB criteria is determining the break points, or the critical corner angle. This study uses a first-order approximation of a feedforward servo system as the system model to complete interpolation and motion control in one sample time. As a result, the tracking error e may be stated as [24].

$$e = F\frac{1-k_f}{k_p} \tag{12}$$

as illustrated in **Figure 3**, where F is the feedrate, k_f is the feedforward gain, and k_p is the position gain.

$\overline{P_1P_2}$ and P_2P_3 are two linked linear blocks with a corner angle y. On P_1P_2, its tracking error is e. Corner errors will occur when the cutting tool goes down the straight lines $\overline{P_1P_2}$ and P_2P_3, beginning at point A. The dashed line represents the actual tool path. Huang [24] presented an isosceles triangle approach to limit the maximum corner error ε_{max}. When considering the geometrical connection of the isosceles triangle ΔAP_2B, the height of the isosceles triangle P_2D reflects the upper bound of the maximum corner error ε_{max}, which is given by Eq. (13)

$$\frac{\varepsilon_{max}}{e} = \cos\frac{\theta}{2} \tag{13}$$

The connections between the prescribed feedrate F_{NC}, the maximum corner error ε_{max}, and the critical corner angle $\theta_{critical}$ are derived by subtracting e from Eqs. (12) and (13).

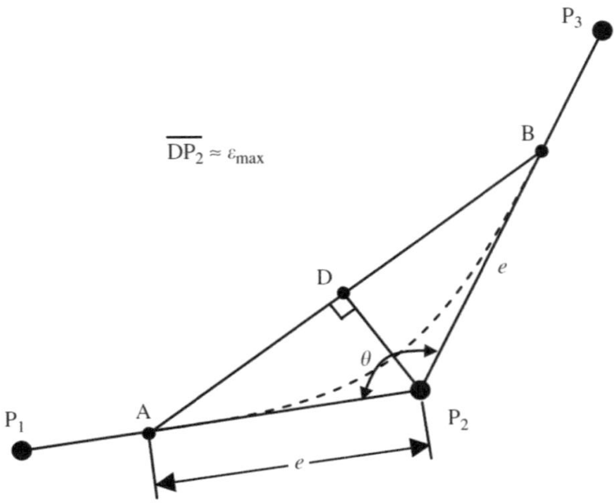

Figure 3.
An isosceles triangle approach is depicted schematically.

$$\theta_{critical} = 2\cos^{-1}\left(\frac{k_p}{1-k_f}\frac{\varepsilon_{max}}{F_{NC}}\right) \tag{14}$$

The FBI's interpolation technique is demonstrated using five CSBs. $\overline{P_0P_1}$ is the initial block in **Figure 4**, $\overline{P_1P_2} - \overline{P_3P_4}$ are the intermediate blocks, and $\overline{P_{41}P_5}$ is the last block. We'll start with the center blocks. Because the interpolated block is $\overline{P_2P_3}$, the basis segment is made up of $\overline{P_1P_2}$, $\overline{P_2P_3}$, and $\overline{P_3P_4}$, as seen in **Figure 4b**. As a result, the FBI may interpolate a Bezier curve $\overline{P_2P_3}$ using a control polygon $Q_0Q_1Q_2Q_3$ and parameters (t_1 and t_2). $\overline{P_1P_2}$, $\overline{P_2P_3}$, and $\overline{P_3P_4}$ are the three middle blocks that have distinct control polygons and may be interpolated into three separate Bezier curve segments, $\overline{P_1P_2}$, $\overline{P_2P_3}$, and $\overline{P_3P_4}$.

4.2. Quintic triangular Bézier patch using dataset's virtual mesh

Liu and Mann [25] introduced a piecewise interpolation polynomial by producing a quintic triangular Bézier patch for each data point using a Bézier control point based on the virtual mesh of the provided dataset. The resulting surface, however, only obtained G^1.

A triangular Bezier patch of degree n is given by

$$K(u,v,w) = \sum_{i+j+k=n} P_{i,j,k}B^n_{i,j,k}(u,v,w) \tag{15}$$

$$B^n_{i,j,k}(u,v,w) = \frac{n!}{i!j!k!}u^iv^jw^k \tag{16}$$

(u, v, w) are barycentric coordinates, and the control points are $P_{i,j,k}$. Liu and Mann [25] constructed a piecewise triangular surface that interpolates data vertices onto a

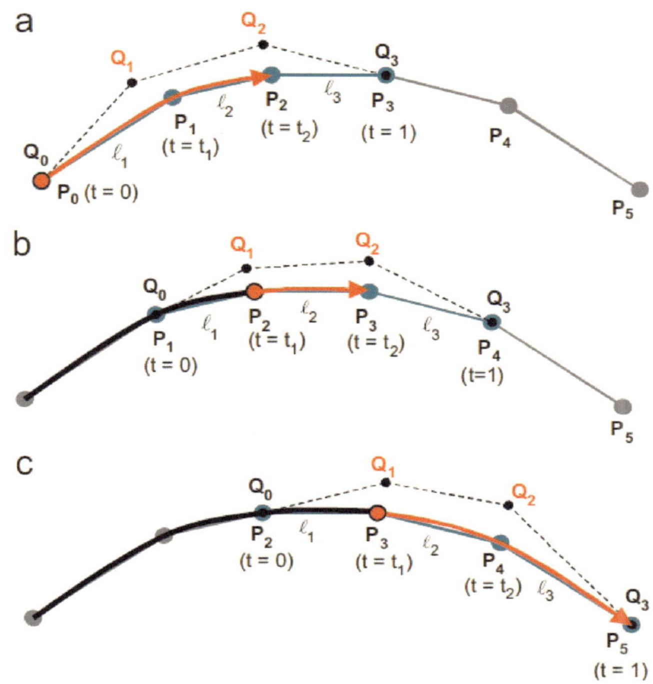

Figure 4.
The FBI interpolates five CSBs: (a) the first and second CSB blocks; (b) the third CSB block; and (c) the fourth and final CSB blocks.

specified triangular mesh M. The number of incident edges for each data vertex V is referred to as the valence of V. Because they assume M has arbitrary topology and is triangulated without singularities, $V's$ valence is always greater than 2. The mesh M is likewise considered to be closed in their study.

Several conditions must be met in order for the triangular patches to connect with G^1 continuity, that is, for the tangent planes from the two adjoining patches to be coplanar at any point along the boundary curve. **Figure 5** depicts two adjoining patches, F_i and F_{i-1}, and their domain triangles for a vertex V with a valence of n.

In domain triangles, the direction of the partial derivative along the ith edge of vertex V is defined as $\vec{u}_i$. $H_i(t)$ is the common boundary curve of patches F_i and F_{i-1}, with $H_i(0) = V$ and $H_i(1) = V_i$. If and only if there are scalar-valued functions $\beta(t), \rho(t)$, and $\tau(t)$ such that

$$\beta_i(t)\frac{\partial F_i}{\partial u_i} = \rho_i(t)\frac{\partial F_i}{\partial u_{i+1}} + \tau_i(t)\frac{\partial F_{i-1}}{\partial u_{i-1}}. \tag{17}$$

F_i and F_{i-1} will join with G^1 continuity along $H_i(t)$.

To ensure that the tangents are properly oriented, we assume $\rho(t)\tau(t) \geq 0$.

Eq. (18) must hold for each pair of neighboring patches in order for patches to meet around vertex V and connect with G^1 continuity. When Eq. (16) is differentiated in the direction of ui and evaluated at $t = 0$, the following equation results:

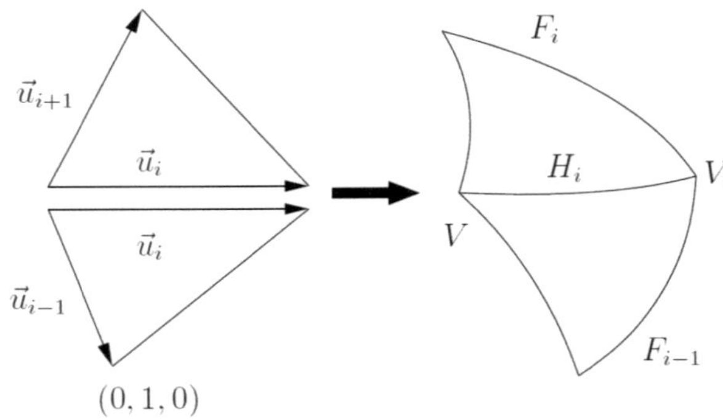

Figure 5.
Adjacent patches.

$$\beta_i'(0)\frac{\partial F_i}{\partial u_i} + \beta_i(0)\frac{\partial^2 F_i}{\partial {u_i}^2} = \beta_i'(0)\frac{\partial F_i}{\partial u_{i+1}} + \beta_i(0)\frac{\partial^2 F_i}{\partial u_{i+1}\partial u_i} + v_i'(0)\frac{\partial F_{i-1}}{\partial u_{i-1}} + v_i(0)\frac{\partial^2 F_{i-1}}{\partial u_{i-1}\partial u_i} \tag{18}$$

By creating n patches around vertex V, a set of equations from Eq. (20) must hold in order for G^1 continuity to exist along all borders; this is known as the twist compatibility or vertex consistency issue [26, 27]. In general, solving these equations necessitates solving circulant matrix problems [28]. Loop's technique constructs boundary curves in such a way that the solutions to the twist terms are ensured [26]. The following are the steps of Loop's sextic scheme:

1. Form quartic boundary polynomials.

2. Go through the twist phrases.

3. Create tangent fields around the edges.

4. Raise the border curve degree to sextic.

5. In the sextic patch, add the second row of control points to interpolate the tangent fields.

6. After averaging the control points from the previous stages, determine the remaining central control point (**Figure 6**).

When the valences of the three vertices of a data triangle are identical, the patch formed by Loop's method is quintic; if the three valences are all six, the patch degree decreases to quartic [26]. The boundary curves in Liu and Mann's [25] method are constructed similarly to the first step of Loop's scheme, but with the center control point set differently. The twist terms are solved in their scheme in the same way as Loop's scheme is solved in the second stage. The first step will be reviewed briefly in this article; the specifics of the remaining parts of Loop's system may be found in [26]. Loop produces the first two control points for each quartic boundary curve with control points of $H_i^0, \ldots, H_i^4$ as

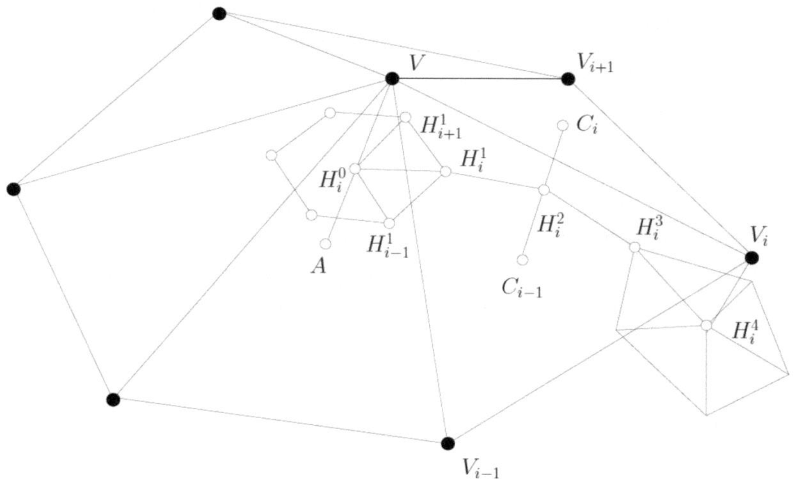

Figure 6.
Tangent boundary construction.

$$H_i^0 = \mu V + (1-\mu)A = \mu V + \frac{1-\mu}{n}\sum_{j=1}^{n} V_j \tag{19}$$

$$H_i^1 = H_i^0 + \frac{\mu}{n}\sum_{j=1}^{n}\left[\cos\left(\frac{2(j-i)\pi}{n}\right)\right]V_j \tag{20}$$

Here, n is the valence of the vertex V, and A is the centroid of all of V's adjacent vertices. The generation of these control points involves performing a first-order Fourier transformation on all of V's surrounding vertices [28]. If all of the control points H_i^0 and H_i^1 are connected, they will create a normal $n-$gon with V as the center, as illustrated in **Figure 2**. μ and η are two shape parameters. When μ is equal to 1, the generated surface interpolates the data vertices. The tangent length at V is defined by the parameter η. Loop proposes that μ and η be used to create an appealing surface form.

$$\mu = \frac{1}{9}\left[4 + \cos\left(\frac{2\pi}{n}\right)\right], \eta = \frac{1}{3}\left[1 + \cos\left(\frac{2\pi}{n}\right)\right]. \tag{21}$$

Eq. (22) calculates the final two control points H_i^3 and H_i^4 from the adjoining vertex V_i. The average of the two centroid points C_i and C_{i+1} of the two neighboring data triangles is used to get the middle point H_i^2:

$$H_i^2 = \frac{C_{i-1} + C_i}{2} = \frac{V}{3} + \frac{V_i}{3} + \frac{V_{i+1}}{6} + \frac{V_{i-1}}{6} \tag{22}$$

4.3. Quadtree Bézier interpolation using uniformly distributed control points

Quadtree decomposition and Bézier interpolation have demonstrated a good ability to extract the image's salient features in infrared and visual image fusion applications. Zhang et al. [29] have indeed developed an iterative quadtree decomposition and Bézier interpolation-based infrared and visual image fusion method. Each image patch is reconstructed using their approach by interpolating its four-by-four uniformly distributed control points, as seen below.

$$q_{i,j}(u,v) = UMP_{i,j}M^TV^T, \tag{23}$$

where (u, v) signifies the data point's position and is represented by the ratio ranging from 0 to 1, (U, V) symbolizes the position-related with coefficient, and M represents the matrix of constant coefficients. $q_{i,j}$ represents the smoothed in image patch, and $P_{i,j}$ indicates the gray levels of 4×4 uniformly distributed control points of $p_{i,j}$.

To be more specific

$$U = [v^3, v^2, v^1, v^0], V = [\nu^3, \nu^2, \nu^1, \nu^0], M = \begin{pmatrix} -1 & 3 & -3 & 1 \\ 3 & -6 & 3 & 0 \\ -3 & 3 & 0 & 0 \\ 1 & 0 & 0 & 0 \end{pmatrix} \tag{24}$$

They used the Bézier interpolation approach to recreate each image patch twice in order to successfully extract the locally distributed bright and dark characteristics. That is, each image patch in the quadtree structure is built first by interpolating the local lowest gray values of the four-by-four uniformly distributed control points, and then a major portion of the bright features are eliminated from the reconstructed picture.

Second, each image patch in the quadtree structure is built further by interpolating the local maximum gray values of the four-by-four uniformly distributed control points; this allows the dark features to be eliminated from the reconstructed picture. As a result, the picture's bright and dark characteristics may be successfully retrieved from the difference image of the original image and the smoothed image.

4.4. Tractable nonlinear interpolation framework using Bézier curves

Shimagaki and Barton [30] proposed a tractable nonlinear interpolation framework using Bézier curves. In addition to incorporating nonlinearity, this approach has the added advantage of conserving sums of categorical variables, which is not guaranteed under arbitrary nonlinear transformations of data. This property can be especially useful for conserved quantities such as probabilities. Historically, the Bézier method has been used in computer graphics to draw smooth curves.

Suppose a polynomial $p(t)$ that is sampled at discrete times t_s for $s \in \{0, 1, \ldots, K\}$. Thus, between two subsequent discrete time points t_s and t_{s+1}, the interpolated value of the polynomial $p_B^{(s)}(t)$ is given by

$$p_B^{(s)}(t) = \sum_{n=0}^{P} \beta_n\left(\frac{t - t_s}{t_{s-1} - t_s}\right)\varphi_n^{(s)}\left(p(t_{s'})_{s'=0}^{K}\right) \tag{25}$$

where β_n is the nth Bernstein polynomial of degree P, with $\beta_n(\theta) = C_n^P\theta^n(1 - \theta)^{P-n} \geq 0$. The control points $\varphi_n^{(s)}\left(p(t_{s'})_{s'=0}^{K}\right)$ is determined by the ensemble of data points $p(t_{s'})_{s'=0}^{K}$ and defines the shape of the curve.

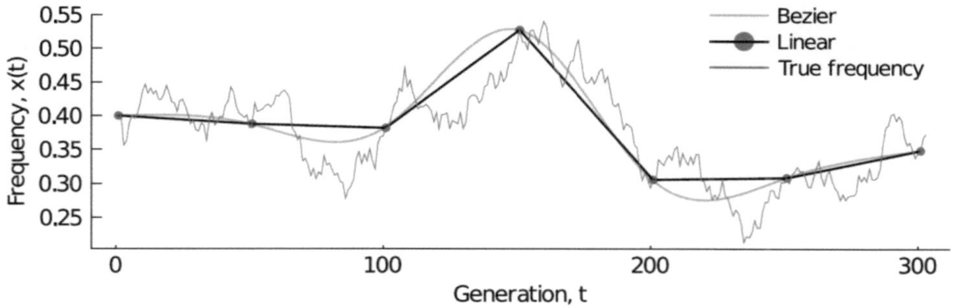

Figure 7.
Smooth curves are generated using Bézier interpolation.

When choosing cubic (P = 3) interpolation for simplicity, but their technique may be extended to polynomials of varying degrees P. To ensure that the section at each interval $[t_s, t_{s+1}]$ for all k is smoothly joined, we apply the following conditions:

$$\varphi_0^{(s)}\left([p(t_{s'})]_{s'=0}^{K}\right) = p(t_s), \tag{26}$$

$$\varphi_3^{(s)}\left([p(t_{s'})]_{s'=0}^{K}\right) = p(t_{s+1}). \tag{27}$$

The other control points $\left\{\varphi_1^{(s)}, \varphi_2^{(s)}\right\}_{s=0}^{K-1}$ are reached by solving an optimization problem that represents the curves' continuity and smoothness restrictions, For example, in **Figure 7**, cubic Bézier curves seamlessly interpolate between discretely sampled frequency trajectories generated by a Wright-Fisher model. Simulation parameters. $L = 50$ sites, $N = 10^3$ population size, mutation rate $\mu = 10^{-3}$, with simulations spanning $T = 300$ generations. Data points are collected every 50 generations and interpolated using cubic Bézier and linear interpolation.

5. Conclusions

In this chapter, unique and inventive methods for maximizing the benefits of various parametric polynomials are discussed. The four new recent polynomials are validated using various interpolation methods. As compared to existing methodologies and conventional procedures, the simulation properties clearly suggest that the examined polynomials may improve the quality of many applications. Additionally, the reviewed method may be used for a variety of image interpolation applications, including image zooming and rotation.

Author details

Mustafa Abbas Fadhel
Mathematics Department, College of Education for Pure Sciences, University of Al Muthanna, Samawa, Iraq

*Address all correspondence to: mustafa@mu.edu.iq

References

[1] Adcock RJ. Note on the method of least squares. The Analyst, Des Moines, Iowa. 1877;**4**:183-184

[2] ISO 10360-6. Geometrical Product Specifications (GPS)—Acceptance and Reverification Test for Coordinate Measuring Machines (CMM)—Part 6: Estimation of Errors in Computing Gaussian Associated Features. Geneva, Switzerland: ISO; 2001

[3] Pearson K. On lines and planes of closest fit to systems of points in space. The Philisophy Magazine. 1991;**2**(11): 559-572

[4] Ahn SJ, Rauh W, Cho HS, Warnecke H-J. Orthogonal distance fitting of implicit curves and surfaces. IEEE Transactions on Pattern Analytical Machine Intellectual. 2002;**24**(5):620-638

[5] Ahn SJ, Westkämper E, Rauh W. Orthogonal distance fitting of parametric curves and surfaces. In: Levesley J et al., editors. Proc. 4th Int'l Symp. Algorithms for Approximation. UK: Univ. of Huddersfield; 2002. pp. 122-129

[6] Boggs PT, Byrd RH, Schnabel RB. A stable and efficient algorithm for nonlinear orthogonal distance regression. SIAM Journal of Science Statistical Computing. 1987;**8**:1052-1078

[7] Boggs PT, Donaldson JR, Byrd RH, Schnabel RB. Algorithm 676 – ODRPACK: Software for weighted orthogonal distance regression. ACM Transactions on Mathematical Software. 1989;**15**(4):348-364

[8] Cao X, Shrikhande N, Hu G. Approximate orthogonal distance regression method for fitting quadric surfaces to range data. Pattern Recognition Letters. 1994;**15**(8):781-796

[9] Helfrich H-P, Zwick D. A trust region method for implicit orthogonal distance regression. Numerical Algorithms. 1993; **5**:535-545

[10] Helfrich H-P, Zwick D. A trust region algorithm for parametric curve and surface fitting. Journal of Computational and Applied Mathematics. 1996;**73**:119-134

[11] Sullivan S, Sandford L, Ponce J. Using geometric distance fits for 3-D object Modeling and recognition. IEEE Transactions on Pattern Analytical Machine. Intellectual. 1994;**16**(12): 1183-1196

[12] Butler BP, Forbes AB, Harris PM. Algorithms for Geometric Tolerance Assessment. NPL: Teddington, U.K; 1994

[13] Sourlier D. Three Dimensional Feature Independent Bestfit in Coordinate Metrology. Zurich, Switzerland: ETH Zurich; 1995

[14] Bookstein FL. Fitting conic sections to scattered data. Computer Graphical Image Processing. 1979;**9**(1):56-71

[15] Fitzgibbon A, Pilu M, Fisher RB. Direct Least Square fitting of ellipses. IEEE Transactions on Pattern Analytical Machine Intellectual. 1999;**21**(5): 476-480

[16] Marshall D, Lukacs G, Martin R. Robust segmentation of primitives from range data in the presence of geometric degeneracy. IEEE Transactions on Pattern Analytical Machine Intellectual. 2001;**23**(3):304-314

[17] Solina F, Bajcsy R. Recovery of parametric models from range images: The case for Superquadrics with global

deformations. IEEE Transactions on Pattern Analytical Machine Intellectual. 1990;**12**(2):131-147

[18] Taubin G. Estimation of planar curves, surfaces, nonplanar space curves defined by implicit equations with applications to edge and range image segmentation. IEEE Transactions on Pattern Analytical Machine Intellectual. 1991;**13**(11):1115-1138

[19] Ameer M, Abbas M, Abdeljawad T, Nazir T. A novel generalization of Bézier-like curves and surfaces with shape parameters. Mathematics. 2022; **10**(3):376

[20] Khan A, Iliyas M, Khan K, Mursaleen M. Approximation of conic sections by weighted Lupaş post-quantum Bézier curves. Demonstratio Mathematica. 2022;**55**(1):328-342

[21] Özger F. On new Bézier bases with Schurer polynomials and corresponding results in approximation theory. Communications Faculty of Sciences University of Ankara Series A1 Mathematics and Statistics. 2020;**69**(1): 376-393

[22] Lepot M, Aubin JB, Clemens F. Interpolation in time series: An introductive overview of existing methods, their performance criteria and uncertainty assessment. Water. 2017; **9**(10):796

[23] Yau H-T, Wang J-B. Fast Bezier interpolator with real-time Lookahead function for high-accuracy machining. International Journal of Machine Tools and Manufacture. 2007;**47**(10):1518-1529

[24] Huang YS. The Research on Acceleration/Deceleration Algorithms for CNC Motion Control. Taiwan: Department of Mechanical Engineering, National Chung Cheng University; 1996

[25] Liu Y, Mann S. Parametric triangular Bézier surface interpolation with approximate continuity. In: Proceedings of the 2008 ACM Symposium on Solid and Physical Modeling. New York, NY: ACM Press; 2008. pp. 381-387

[26] Loop C. A G^1 triangular spline surface of arbitrary topological type. Computer Aided Geometric Design. 1994;***11***(3):303-330

[27] Van Wijk J. Bicubic patches for approximating nonrectangular control-point meshes. Computer Aided Geometric Design. 1986;***3***(1):1-13

[28] Davis P. Circulant Matrices. New York: Wiley; 1979

[29] Zhang Y, Zhang L, Shen J, Zhang S, Bai X. Infrared and visual image fusion via iterative quadtree decomposition and Bézier interpolation. In: Fourteenth International Conference on Digital Image Processing (ICDIP 2022). Vol. 12342. Wuhan, China: SPIE; 2022. pp. 730-738

[30] Shimagaki K, Barton JP. Bézier interpolation improves the inference of dynamical models from data. Physical Review E. 2023;**107**(2):024116

Chapter 5

Rediscovery of Routh Polynomials after Hundred Years in Obscurity

Gregory Natanson

Abstract

The introductory part of the paper outlines misfortunate history of one of Routh's most remarkable works. It points to an important distinction between infinitely many Routh polynomials forming a differential polynomial system (DPS) and their finite orthogonal subset referred to as "Romanovski-Routh" (R-Routh) polynomials (Romanovski/pseudo-Jacobi polynomials in Lesky's classification scheme). It was shown that there are two infinite sequences of Routh polynomials without real roots (in contrast with R-Routh polynomials with all the roots located on the real axis). The factorization functions (FFs) formed by polynomials from these infinite sequences can be thus used to generate exactly solvable rational Darboux transforms of the canonical Sturm-Liouville equation (CSLE) quantized in terms of R-Routh polynomials with degree-dependent indexes. The current analysis is focused solely on the CSLE with the density function having two simple zeros on the opposite sides of the imaginary axis. A special attention is given to the limiting case of the density function with the simple poles at $\pm i$ when the given CSLE becomes translationally form-invariant, and as a result, the eigenfunctions of its rational Darboux transforms are expressible in terms of a finite number of exceptional orthogonal polynomials (EOPs).

Keywords: differential polynomial system, Routh polynomials, pseudo-Jacobi polynomials, Romanovski-Routh polynomials, rational canonical Sturm-Liouville equation, rational Darboux transform, Liouville-Darboux transformations, exceptional orthogonal polynomials

1. Introduction

Let us first outline some groundbreaking aspects of Routh' paper [1] overlooked by mathematicians for more than a hundred years before a brief reference to his results appeared in Ismail's monograph [2]. Routh' precocious discovery not properly appreciated even today (see [3] for more detailed critical remarks) was the classification of the *real* second-order differential eigenequations of hypergeometric type

$$\sigma(x)\frac{d^2X_n}{dx^2} + \tau(x)\frac{dX_n}{dx} + h_nX_n(x) = 0 \tag{1}$$

according to positions of zeros of the leading polynomial coefficient

$$\sigma(x) \equiv ax^2 + bx + c, \tag{2}$$

regardless of the choice of the polynomial coefficient of the first derivative:

$$\tau(x) \equiv fx + g. \tag{3}$$

Namely Routh proved that the function

$$X_n(x) = \frac{\sigma(x)}{\mathrm{R}(x)} \frac{d^n}{d^n x} \left[\sigma^{n-1}(x)\mathrm{R}(x)\right] \tag{4}$$

where R(x) is a solution of the first-order ordinary differential equation (ODE)

$$\frac{1}{\mathrm{R}(x)} \frac{d\mathrm{R}}{dx} = \frac{\tau(x)}{\sigma(x)} \tag{5}$$

(see his Art. 18, coupled with Eq. (6) in Art. 4) obeys eigeneq. (1) for

$$h_n = -n\left[f + a(n-1)\right]. \tag{6}$$

(Regretfully neither Rodriguez' [4] nor Jacobi's [5] classic works were mentioned in this connection.) Fifty years later, Hildebrandt [6] started from the reminiscent first-order ODE

$$\frac{1}{\omega(x)} \frac{d\omega}{dx} = \frac{N(x)}{D(x)} \tag{7}$$

referring to it as the "*Pearson differential equation*" [7] and then used its analytical solutions as the weight functions for the generalized Rodrigues formula [4, 5].

$$X_n(x) = \frac{1}{\omega(x)} \frac{d^n}{d^n x} [D^n(x)\omega(x)], \tag{8}$$

where $D(x) \equiv \sigma(x)$ and

$$N(x) \equiv \tau(x) - \sigma'(x) \tag{9}$$

in terms of [8]. Hildebrandt then proved that the polynomials generated via (8) must satisfy second-order differential eigenequation (1). Comparing (7) and (9) with (5) thus gives

$$\mathrm{R}(x) = \sigma(x)\omega(x) \tag{10}$$

so Routh's formula (4) is nothing but the *precursor* of (8).

Examination of all possible solutions of first-order ODE (7) brought Hildebrandt to the classification of various types of polynomial solutions of second-order differential eigenequation (1) including different sub-cases of five cases specified by Routh:

i. discriminant of quadratic polynomial (2) is positive;

ii. discriminant of quadratic polynomial (2) is negative;

iii. quadratic polynomial (2) has a double root;

iv. the leading coefficient function is a polynomial of the first degree;

v. the leading coefficient function is a constant.

Combining subcases i) and i') into a single class of leading polynomial coefficient with a nonzero discriminant, we come to Bochner's [9] four classes of second-order differential eigenequations (1) defined over complex field (i.e., Jacobi, Bessel, Laguerre, and Hermite polynomials accordingly). While case i) is nothing but the notorious Jacobi polynomials [5] leading coefficient function (2) with two complex-conjugated zeros is associated with a completely different *infinite* real polynomial sequence termed "Routh polynomials" below.

In following Everitt et al. [10, 11], we refer to Routh's cases i) – v) as differential polynomial systems (DPSs), while preserving the term "orthogonal polynomial system" (OPS) used by Kwon and Littlejohn [12] in this context solely for classical orthogonal polynomials forming infinite sequences of *positive definite* orthogonal polynomials. As stressed by Kwon and Littlejohn [12], Bochner himself "did not mention the orthogonality of the polynomial systems that he found. The problem of classifying all classical orthogonal polynomials was handled by many authors thereafter" based on his analysis of possible polynomial solutions of *complex* second-order differential eigenequations.

The crucial point is that polynomial solutions of the differential eigenequation

$$(x^2+1)\frac{d^2X_n}{dx^2}+(fx+g)\frac{dX_n}{dx}+h_n(f,g)X_n(x)=0 \tag{11}$$

referred to below as Routh polynomials represent a supplementary *nontrivial* real-field reduction of the complex Jacobi DPS, in addition to conventional *real* Jacobi polynomials. Obviously, this assertion directly follows from Bochner's renowned paper [9]; however, it took another 40 years before Cryer [13] came up with the well-known [2] representation of Routh polynomials as Jacobi polynomials with complex conjugated indexes in an imaginary argument (see [3] for more details). Namely Cryer realized that the weight function for Jacobi polynomials:

$$w^{(\alpha,\beta)}(\eta)\equiv(1-\eta)^{\alpha}(1+\eta)^{\beta} \tag{12}$$

formally coincides with Pearson's distribution function of type IV [7]:

$$w^{(\alpha,\alpha*)}(ix)=w^{(\alpha_R+i\alpha_I)}(x)\equiv\left(1+x^2\right)^{\alpha_R}exp\,(2\alpha_I\,arctan\,x), \tag{13}$$

where α_R and α_I are real and imaginary parts of the complex number α. This brought him to the renowned formula

$$\mathfrak{R}_m^{(\alpha_R+i\,\alpha_I)}[x]\equiv(-i)^m P_m^{(\alpha_R+i\,\alpha_I,\alpha_R-i\,\alpha_I)}(ix), \tag{14}$$

where

$$\mathfrak{R}_m^{(\alpha_R+i\,\alpha_I)}[x]\equiv P_m(x;\alpha_R,\alpha_I) \tag{15}$$

in Ismail's notation [2]. As discussed more thoroughly in [3], the monic polynomials

$$\hat{\mathfrak{R}}_m^{(\alpha_R+i\,\alpha_I)}[x] = \frac{1}{K_m(\alpha_R)\,w^{(\alpha_R+i\alpha_I)}(x)}\frac{d^m}{d^m x}\left[\left(x^2+1\right)^m w^{(\alpha_R+i\alpha_I)}(x)\right] \tag{16}$$

introduced by Cryer with

$$K_m(\alpha_R) \equiv \frac{(2\alpha_R+2m)_m}{m!2^m} \tag{17}$$

coincide with the polynomials $P_m(x;\nu,N)$ in [14] with N = $-\alpha_R - 1$, $\nu = \alpha_I$:

$$\hat{\mathfrak{R}}_m^{(\alpha_R+i\,\alpha_I)}[x] \equiv P_m(x;\alpha_I,-\alpha_R-1). \tag{18}$$

In this paper, we prefer to adopt the notation of the cited monograph [14] to make use of the formulas listed in §9.9 for the so-called "pseudo-Jacobi polynomials," with N changed for an arbitrary *real* number. (A similar suggestion has been already put forward in [15] though with the lower bound of $-1/2$ for N.) Note that the term "Routh polynomials" is used by us in exactly the same sense as "pseudo-Jacobi polynomials" in [16], that is, without any limitations on sign and values of α_R. The finite orthogonal subsequence of the Routh DPS discovered by Romanovsky [17][1] is referred to by us as "Romanovski-Routh" (R-Routh) polynomials similarly to the epithet "Romanovski/pseudo-Jacobi" suggested for these polynomials by Lesky [18, 19].

2. Rational Sturm-Liouville problem solvable via R-Routh polynomials with degree-dependent indexes

Author's own interest in Routh polynomials was stimulated by examination of the Routh-reference (RRef) canonical Sturm-Liouville equation (CSLE) [20, 21].

$$\left\{\frac{d^2}{d\eta^2} + {}_iI[\eta;h_o;\kappa;\varepsilon]\right\}{}_i\Phi[\eta;h_o;\kappa;\varepsilon] = 0, \tag{19}$$

where the real "Bose invariant" [22].

$${}_iI[\eta;h_o;\kappa;\varepsilon] \equiv {}_iI^o[\eta;h_o] + \varepsilon_i\rho\,[\eta;\kappa] \tag{20}$$

in Milson's terms [20] represents a superposition of the reference polynomial fraction (RefPF)

$${}_iI^o\,[\eta;h_o] = -\frac{h_o}{4(\eta+i)^2} - \frac{h_o^*}{4(\eta-i)^2} + \frac{2h_{o;R}+1}{4(\eta^2+1)} \tag{21}$$

$$= \frac{h_{o;R} - h_{o;I}\,\eta}{(\eta^2+1)^2} + \frac{1}{4(\eta^2+1)} \tag{22}$$

---

[1] In all the relevant papers other than [17], Vsevolod Romanovskii spelled his name as "Romanovsky," so we use the latter spelling when mentioning him by name.

and the density function

$$ {}_{i}\rho\,[\eta;\kappa] = \frac{T_2[\eta;\kappa]}{(\eta^2+1)^2} \tag{23}$$

dependent accordingly on a complex parameter

$$ h_o \equiv h_{o;R} + i\,h_{o;I} \tag{24}$$

and on the second-degree monic "tangent polynomial" (TP)

$$ {}_{i}T_2[\eta;\kappa] \equiv \eta^2 + \kappa \equiv \left(\eta - i\sqrt{\kappa}\right)\left(\eta + i\sqrt{\kappa}\right) \tag{25}$$

with the negative discriminant -κ. The energy reference point for RCSLE (19) was chosen in such a way that the indicial equation for the pole at infinity has real (complex-conjugated) roots at any negative (any positive) energy. The density function with a constant TP in numerator, giving rise to the translationally form-invariant (TFI) CSLE of Group B with the trigonometric Liouville potential [23], requires a special consideration.

It has been proven in [21] that the eigenfunctions of CSLE (19) can be expressed in terms of R-Routh polynomials:

$$ {}_{i}\phi_n[\eta; h_{o;R} + i h_{o;I};\kappa] \propto (1 - i\eta)^{{}_{i}\rho_n}(1 + i\eta)^{{}_{i}\rho_n^{*}}\, R_n^{(2\,\lambda_{n;I},\lambda_{n;R}+1)}(\eta) \tag{26}$$

with

$$ {}_{i}\lambda_n \equiv \lambda_{n;R} + i\,\lambda_{n;I} \equiv 2\,{}_{i}\rho_n - 1 \tag{27}$$

which are defined via (46) and (47) in [24]:[2]

$$ R_n^{(2\alpha_I,\alpha_R+1)}(\eta) = (-i)^n\, P_n^{(\alpha_R+i\alpha_I,\alpha_R-i\alpha_I)}(i\eta) \equiv \mathfrak{R}_n^{(\alpha_R+i\alpha_I)}(\eta) \tag{28}$$

for

$$ n \le N_0 \equiv \lfloor -\alpha_R - 1/2 \rfloor . \tag{29}$$

One can easily verify that eigenfunctions (26) are square integrable with density function (23) as far as condition (29) holds. To prove that the Sturm-Liouville problem in question is *exactly* solvable, the author [21] took advantage of Stevenson's idea [27] to express an analytically continued solution in terms of hypergeometric polynomials in complex argument. It was just confirmed that the latter *formally complex* polynomials can be converted into real R-Routh polynomials (28).

Examination of the exponent differences for three (including ∞) poles of CSLE (19) reveals that the sought-for real parameters $\lambda_{n;R}$ and $\lambda_{n;I}$ are unambiguously determined by the following set of the algebraic equations [21]:

$$ {}_{i}\lambda_n^2 = h_{o;R} + i\,h_{o;I} + 1 + (1-\kappa)\,{}_{i}\rho_{\infty;n}^2 \tag{30}$$

[2] Note that Quesne [24] slightly modified the notation for the Romanovski polynomials compared with [25, 26]. We prefer to follow her notation to be able to match our formulas for the limiting case κ = 1 to her results for the rational Darboux transform (RDT) of the "Scarf II" potential.

and

$$ {}_{i}\rho_{\infty;n} = -\lambda_{n;R} - n - 1/2 > 0, \tag{31} $$

with the square of real parameter (31) determining the magnitude of the corresponding eigenvalue

$$ {}_{i}\varepsilon_{n}(h_o;\kappa) = -{}_{i}\rho^{2}_{\infty;n} < 0. \tag{32} $$

Keeping in mind that

$$ 2\lambda_{n;R}\lambda_{n;I} = h_{o;I} \tag{33} $$

we come to the quartic eq. [21]

$$ \lambda^{4}_{n;R} - [h_{o;R} + 1 + (1-\kappa)(\lambda_{n;R} + n + 1/2)]\lambda^{2}_{n;R} - 1/4 h^{2}_{o;I} = 0 \tag{34} $$

with the positive leading coefficient κ and the negative free term (except the limiting case $h_{o;I} = 0$ [28] associated with the symmetric Ginocchio potential [29]). This implies that quartic eq. (34) necessarily has at least two real roots $\lambda_{n;R} \equiv \lambda_{\mathbf{c},n;R} < 0$ and $\lambda_{\mathbf{d},n;R} > 0$ if CSLE (19) has at minimum n+1 discrete energy levels. We thus assert that each eigenfunction (26) must be accompanied by the second quasi-rational solution (q-RS)

$$ {}_{i}\phi_{\mathbf{d},n}[\eta;h_o;\kappa] = (1+i\eta)^{{}_{i}\rho_{\mathbf{d},n}}(1-i\eta)^{{}_{i}\rho^{*}_{\mathbf{d},n}}\mathfrak{R}_{n}^{({}_{i}\lambda_{\mathbf{d},n})}(\eta), \tag{35} $$

$$ \equiv (1+\eta^{2})^{1/2(\lambda_{\mathbf{d},n;R}+1)} exp\,(1/2\lambda_{\mathbf{d},n;I}\ arctan\ \eta)\mathfrak{R}_{n}^{({}_{i}\lambda_{\mathbf{d},n})}(\eta) \tag{36} $$

at the energy

$$ {}_{i}\varepsilon_{\mathbf{d},n} = -{}_{i}\rho^{2}_{\infty;\mathbf{d},n}, \tag{37} $$

where we set

$$ {}_{i}\lambda_{\mathbf{d},n} = \lambda_{\mathbf{d},n;R} + 1/2(h_{o;I}/\lambda_{\mathbf{d},n;R})\,i, \tag{38} $$

$$ {}_{i}\rho_{\mathbf{d},n} = 1/2({}_{i}\lambda_{\mathbf{d},n} + 1), \tag{39} $$

and

$$ {}_{i}\rho_{\infty;\mathbf{d},n} = -\lambda_{\mathbf{d},\mathrm{n};R} - n - 1. \tag{40} $$

Here, in following the classification of the "factorization functions" (FFs) suggested by us for Darboux transformations of radial potentials [30] (years before birth of supersymmetric quantum mechanics [31, 32]), we use the labels **c** and **d** to specify the eigenfunctions of the given Sturm-Liouville problem and their counterparts infinite at both ends (type III in Quesne's terms [24]).

My own interest in the q-RSs of type **d** was stimulated by Quesne's conjecture [24] concerning the existence of Routh polynomials with no real roots, which can be thereby used to construct nodeless FFs for rational Darboux transformations giving rise to new exactly solvable rational potentials.

3. Two infinite sequences of nodeless q-RSs composed of Routh polynomials

Rewriting quartic eq. (34) as

$$\lambda^4_{\dagger,m;R} - \left[h_{o;R} + 1 + (1-\kappa)\left(\lambda_{\dagger,m;R} + m + {}^1\!/_2\right)^2 \lambda^2_{\dagger,m;R} - {}^1\!/_4 h^2_{o;I} = 0 \tag{41}$$

and keeping in mind that its leading coefficient κ and free term have opposite signs (again except the mentioned symmetric case), we assert the given quartic equation always has at least two real roots and therefore a pair of unbounded q-RSs

$$_i\phi_{\mathbf{d},\mathrm{m}}[\eta; h_o; \kappa] = \sqrt{\eta^2+1}\,(1+i\eta)^{\frac{1}{2i}\lambda_{\mathbf{d},\mathrm{m}}}(1-i\eta)^{\frac{1}{2i}\lambda^*_{\mathbf{d},\mathrm{m}}}\mathfrak{R}_m^{({}_i\lambda_{\mathbf{d},\mathrm{m}})}[\eta] \text{ with } \lambda_{\mathbf{d},\mathrm{m};\,\mathrm{R}} > 0 \tag{42}$$

and

$$_i\phi_{\dagger',\mathrm{m}}[\eta; h_o; \kappa] = \sqrt{\eta^2+1}\,(1+i\eta)^{\frac{1}{2i}\lambda_{\dagger',\mathrm{m}}}(1-i\eta)^{\frac{1}{2i}\lambda^*_{\dagger',\mathrm{m}}}\mathfrak{R}_m^{({}_i\lambda_{\dagger',\mathrm{m}})}[\eta] \text{ with } \lambda_{\dagger',\mathrm{m};\,\mathrm{R}} < 0 \tag{43}$$

formed by Routh polynomials exists for any nonnegative integer m, with **t'** = **c** or **d'**.

Dividing quartic eq. (41) by m^4, setting

$$C_\dagger = \lim_{m\to\infty}\left(\lambda_{\dagger,m;R}/m\right), \tag{44}$$

and making m tend to ∞, we come to the quadratic equation

$$C^2_\dagger + (\kappa-1)\,(C_\dagger+1)^2 = 0 \tag{45}$$

in $C_\dagger$, which has the positive leading coefficient κ and negative (positive) free term for $0<\kappa<1$ ($\kappa>1$), so its two roots always have opposite signs:

$$C_{\mathbf{d}} > 0, C_{\mathbf{d}'} < 0 \text{ for any } \kappa < 1. \tag{46}$$

One can directly verify that $C_{\mathbf{d}'}$ necessarily differs from -1 and therefore the magnitudes of the intrinsic characteristic exponents (ChExps) at infinity

$$_i\rho_{\infty;\dagger,m} = -\lambda_{\dagger,m;R} - m - 1 < 0 \qquad (\dagger = \mathbf{d} \text{ or } \mathbf{d}') \tag{47}$$

monotonically grow with m for sufficiently large m iff $\kappa<1$. Under the latter condition, the energies of both q-RSs (42) and (43),

$$_i\varepsilon_{\dagger,m} = -{}_i\rho^2_{\infty;\dagger,m}, \tag{48}$$

must thus lie below the ground energy level $_i\varepsilon_{\mathbf{c}0}$ for m > > 1. As a direct consequence of the disconjugacy theorem recently brought into the theory of rationally extended potentials by Grandati et al. [33–37], we conclude that the solutions in question may not have more than one node. Taking into account that solutions of any SLE may have only a simple zero at any regular point and also that the monic polynomial of an even degree is necessarily positive for sufficiently large absolute values of its argument, we conclude that q-RSs (42) and (43) with κ between 0 and 1 must be nodeless if m = 2j > > 1 assuming that leading coefficient (17) of Routh polynomial (14) differs from zero. The latter constraint is always valid for any q-RS (42) since the real part of the index $\lambda_{\mathbf{d},m;R}$ is positive by definition. The cited constraint also holds for q-RS (43) unless 2 $\lambda_{\mathbf{d}',m;R}$ is a negative integer smaller than 1–m.

As discussed in next section, each of infinitely many nodeless q-RSs (42) and (43) can be used as an FF to construct the RDT of CSLE (19) exactly solvable by polynomials. The latter constitute a new type of polynomials originally discovered by the author [38] while analyzing a structure of eigenfunctions for the Cooper-Ginocchio-Khare potential [39]. The distinctive common feature of these polynomials [40] is that they satisfy Heine-type [41, 42] differential equations with the first-derivative coefficient function dependent on the polynomial degree. The polynomial sequences in question turn into finite or infinite sequences of exceptional orthogonal polynomials (EOPs) in Quesne's terms [24] if the exponent differences (ExpDiffs) for the SLE poles in the finite plane become energy-independent [43], and as a result, the corresponding Liouville potential belongs to Group A of translationally shape-invariant (TSI) potentials in Odake and Sasaki's [44, 45] classification scheme. The RDTs of the translationally form-invariant (TFI) CSLEs from Group B [46] represent a more typical case of rational CSLEs (RCSLEs) converted by gauge transformations to Heine-type differential equations with the first-derivative coefficient function dependent on the polynomial degree (see [34, 37, 45, 47, 48] for examples). Again we deal with the polynomial solutions of a new type (referred to by us as Gauss-seed Heine polynomials [40]), which are in general not expressible in terms of EOPs, in contrast with the statement made in [37].

Before going to the discussion of RDTs of CSLE (19), it is convenient to reformulate the corresponding spectral problem by introducing the (*algebraic*) "prime" SLE [49].

$$\left\{\frac{d}{d\eta}\left(\eta^2+1\right)^{\frac{1}{2}}\frac{d}{d\eta}-{}_{i}q\left[\eta;h_o\right]+\varepsilon\,{}_{i}w[\eta;\kappa]\right\}{}_{i}\Psi[\eta;h_o;\kappa;\varepsilon]=0 \tag{49}$$

obtained from CSLE (19) by the gauge transformation

$${}_{i}\Psi[\eta;h_o;\kappa;\varepsilon]=\left(\eta^2+1\right)^{-\frac{1}{4}}{}_{i}\Phi[\eta;h_o;\kappa;\varepsilon] \tag{50}$$

so

$${}_{i}w[\eta;\kappa]\equiv\left(\eta^2+1\right)^{\frac{1}{2}}{}_{i}\rho[\eta;\kappa]. \tag{51}$$

The particular form of the energy-free term ${}_{i}q\left[\eta;h_o\right]$ is nonessential for our discussion. and we refer the reader to [3] for all the details.

The main advantage of converting CSLE (19) to its prime form with respect to the *regular* singular point at infinity comes from our observation [49] that the ChExps for this pole have opposite signs, and therefore, the corresponding principal Frobenius solution is unambiguously selected by the Dirichlet boundary conditions (DBCs):

$$\lim_{\eta\to\pm\infty}{}_{i}\Psi_{\mathbf{c},n}[\eta;h_o;\kappa]=0 \text{ for } \mathrm{n}=0,\ \ldots,\mathrm{n}_{\max} \tag{52}$$

imposed on its eigenfunctions

$$
{}_i\psi_{\mathbf{c},n}[\eta;h_o;\kappa] \equiv {}_i\Psi\left[\eta;h_o;\kappa;{}_i\varepsilon_{\mathbf{c},n}(h_o;\kappa)\right].
$$

$$
= \left(\eta^2+1\right)^{-\frac{1}{4}}{}_i\phi_{\mathbf{c},n}[\eta;h_o;\kappa]. \tag{53}
$$

(Remember that the energy reference point was chosen by us in such a way that the indicial equation for the pole at infinity has real roots at any negative energy.) Reformulating the given spectral problem in such a way allows us to take advantage of powerful theorems proven in [50] for zeros of principal solutions of SLEs solved under the DBCs at singular ends.

4. Exactly solvable rational Liouville-Darboux transforms of RRef CSLE

As discussed in previous section, any q-RS

$$
{}_i\phi_{\dagger,2\mathrm{j}}[\eta;h_o;\kappa] = \sqrt{\eta^2+1}\,(1+i\eta)^{\frac{1}{2}{}_i\lambda_{\dagger,2\mathrm{j}}}(1-i\eta)^{\frac{1}{2}{}_i\lambda^*_{\dagger',2\mathrm{j}}}\mathfrak{R}_{2\mathrm{j}}^{\left({}_i\lambda_{\dagger,2\mathrm{j}}\right)}[\eta] \tag{54}
$$

at the energy

$$
{}_i\varepsilon_{\dagger,2\mathrm{j}}(h_o;\kappa) < {}_i\varepsilon_{\mathbf{c}0}(h_o;\kappa)\ \left(\dagger = \mathbf{d}\text{ or }\mathbf{d}'\right) \tag{55}
$$

is nodeless for $\mathrm{j} >> 1$ unless $2\lambda_{\mathbf{d}',2\mathrm{j};R}$ is a negative integer smaller than 1–2j for $\dagger = \mathbf{d}'$. (In the latter case, the degree of the Routh polynomial is smaller than 2j and may happen to be odd.) Each of these solutions can be thus used as an FF for the rational Liouville-Darboux transformation (RLDT) such that Rudyak and Zahariev's reciprocal function [51].

$$
{}_i\phi_{\mathbf{c},0}[\eta;h_o;\kappa|\dagger,2\mathrm{j}] = \frac{{}_i\rho^{-\frac{1}{2}}[\eta;\kappa]}{{}_i\phi_{\dagger,2\mathrm{j}}[\eta;h_o;\kappa]} \qquad \left(\dagger = \mathbf{d}\text{ or }\mathbf{d}'\right) \tag{56}
$$

represents a q-RS of the transformed RCSLE:

$$
\left\{\frac{d^2}{d\eta^2} + {}_iI^o[\eta;h_o;\kappa|\dagger,2\mathrm{j}] + {}_i\varepsilon_{\dagger,2\mathrm{j}}(h_o;\kappa)\,{}_i\rho[\eta;\kappa]\right\}{}_i\phi_{\mathbf{c},0}[\eta;h_o;\kappa|\dagger,2\mathrm{j}] = 0 \tag{57}
$$

at same energy (55), where

$$
\begin{aligned}
{}_iI^o[\eta;h_o;\kappa|\dagger,2\mathrm{j}] = &-ld^2{}_i\phi_{\mathbf{c},0}[\eta;h_o;\kappa|\dagger,2\mathrm{j}] - \dot{ld}\,{}_i\phi_{\mathbf{c},0}[\eta;h_o;\kappa|\dagger,2\mathrm{j}]\\
&-{}_i\varepsilon_{\dagger,2\mathrm{j}}(h_o;\kappa)\,{}_i\rho[\eta;\kappa]
\end{aligned} \tag{58}
$$

with *ld* and dot standing for the logarithmic derivative and the derivative with respect to η, respectively. It will be proven below that q-RS (56) represents the lowest-energy eigenfunction of the given Sturm-Liouville problem as indicated by its label.

We [40] introduced the term "Liouville-Darboux transformation" (LDT) to stress that we deal with the three-step operation:

i. the Liouville transformation [20, 22] from the given SLE to the conventional Schrödinger equation;

ii. the Darboux deformation of the corresponding Liouville potential;

iii. the inverse Liouville transformation from the Schrödinger equation to the new SLE *preserving* both leading coefficient function and weight.

Similarly to the discussion presented in previous Section 2 for RRef CSLE (19), the gauge transformation

$$ {}_i\Psi[\eta;h_o;\kappa;\varepsilon|\dagger,2\mathrm{j}] = (\eta^2+1)^{-1/4}\,{}_i\Phi[\eta;h_o;\kappa;\varepsilon|\dagger,2\mathrm{j}] \tag{59} $$

converts the CSLE

$$ \left\{\frac{d^2}{d\eta^2} + {}_iI^o[\eta;h_o;\kappa|\dagger,2\mathrm{j}] + \varepsilon_i\rho[\eta;\kappa]\right\}{}_i\Phi[\eta;h_o;\kappa;\varepsilon|\dagger,2\mathrm{j}] = 0 \qquad (\dagger = \mathbf{d}\text{ or }\mathbf{d}') \tag{60} $$

into the prime SLE

$$ \left\{\frac{d}{d\eta}(\eta^2+1)^{\frac{1}{2}}\frac{d}{d\eta} - {}_iq[\eta;h_o|\dagger,2\mathrm{j}] + \varepsilon_i w[\eta;\kappa]\right\}{}_i\Psi[\eta;h_o;\kappa;\varepsilon|\dagger,2\mathrm{j}] = 0 \tag{61} $$

analytically solved under the DBCs

$$ \lim_{\eta\to\pm\infty}{}_i\Psi[\eta;h_o;\kappa;\varepsilon|\dagger,2\mathrm{j}] = 0 \qquad (\dagger = \mathbf{d}\text{ or }\mathbf{d}'). \tag{62} $$

Setting

$$ {}_i\psi_{\dagger,m}[\eta;h_o;\kappa] = (\eta^2+1)^{-1/4}\,{}_i\phi_{\dagger,m}[\eta;h_o;\kappa] \qquad (\dagger = \mathbf{c},\mathbf{d},\ \text{or }\mathbf{d}') \tag{63} $$

we assert that the RLDT in question inserts the new ground energy level associated with the nodeless eigenfunction

$$ {}_i\psi_{\mathbf{c}0}[\eta;h_o;\kappa|\dagger,2\mathrm{j}] = \sqrt{\frac{\eta^2+1}{T_2[\eta;\kappa]}}\,{}_i\psi^{-1}_{\dagger,2\mathrm{j}}[\eta;h_o;\kappa] \tag{64} $$

$$ \propto \eta^{-2\mathrm{j}-1}(1+\eta^2)^{-\frac{1}{2}{}_i\lambda_{\dagger,2\mathrm{j};\mathrm{R}}(h_o;\kappa)+\frac{1}{4}} \quad \text{for } |\eta|>>1 \ (\dagger = \mathbf{d}\text{ or }\mathbf{d}') \tag{65} $$

at the energy

$$ {}_i\varepsilon_{\mathbf{c},0}(h_o;\kappa|\dagger,2\mathrm{j}) = {}_i\varepsilon_{\dagger,2\mathrm{j}}(h_o;\kappa) < {}_i\varepsilon_{\mathbf{c}0}(h_o;\kappa) \ (\dagger = \mathbf{d}\text{ or }\mathbf{d}') \tag{66} $$

for sufficiently large j. (Remember that ${}_i\lambda_{\mathbf{d},2\mathrm{j};\mathrm{R}}(h_o;\kappa)$ is necessarily positive in this case, whereas the function ${}_i\psi_{\mathbf{d}',2\mathrm{j}}[\eta;h_o;\kappa]$ in (64) for $\dagger = \mathbf{d}'$ is not an eigenfunction of the prime SLE *by definition*, so it must infinitely grow as $\eta\to\pm\infty$).

To obtain explicit formula for higher-energy eigenfunctions formed by the RDTs of the eigenfunctions of SLE (49),

$$ {}_i\psi_{\mathbf{c},n}[\eta;h_o;\kappa] = \left(\eta^2+1\right)^{-\frac{1}{4}}\, {}_i\phi_{\mathbf{c},n}[\eta;h_o;\kappa], \tag{67}$$

we take advantage of Rudyak and Zahariev's conventional formula [51]:

$$ \Phi[\eta;h_o;\kappa;\varepsilon|\dagger,2\mathrm{j}] = \frac{W\left\{{}_i\phi_{\dagger,2\mathrm{j}}[\eta;h_o;\kappa],\Phi[\eta;h_o;\kappa;\varepsilon]\right\}}{{}_i\rho^{\frac{1}{2}}[\eta;\kappa]\,{}_i\phi_{\dagger,2\mathrm{j}}[\eta;h_o;\kappa]} \tag{68}$$

for the general solution of the generic CSLE and represent the mentioned RDTs as

$$ \Psi\left[\eta;h_o;\kappa;{}_i\varepsilon_{\mathbf{c},n}(h_o;\kappa)|\dagger,2\mathrm{j}\right] = \frac{W\left\{{}_i\psi_{\dagger,2\mathrm{j}}[\eta;h_o;\kappa],{}_i\psi_{\mathbf{c},n}[\eta;h_o;\kappa]\right\}}{{}_i\rho^{\frac{1}{2}}[\eta;\kappa]\,{}_i\psi_{\dagger,2\mathrm{j}}[\eta;h_o;\kappa]}$$

$$ \text{for } \mathrm{n}=0,\ \dots,\ \mathrm{n_{max}}\ \left(\dagger = \mathbf{d} \text{ or } \mathbf{d}'\right) \tag{69}$$

or, which is equivalent,

$$ \begin{aligned} \Psi\left[\eta;h_o;\kappa;{}_i\varepsilon_{\mathbf{c},n}(h_o;\kappa)|\dagger,2\mathrm{j}\right] &= {}_i\rho^{-\frac{1}{2}}[\eta;\kappa]\,{}_i\psi_{\mathbf{c},n}[\eta;h_o;\kappa] \\ &\times\left\{ld_i\psi_{\mathbf{c},n}[\eta;h_o;\kappa] - ld\,{}_i\psi_{\dagger,2\mathrm{j}}[\eta;h_o;\kappa]\right\}. \end{aligned} \tag{70}$$

Keeping in mind that

$$ \lim_{\eta\to\pm\infty}\left|{}_i\rho^{-\frac{1}{2}}[\eta;\kappa]\ ld\,\psi_{\mathbf{c},n}[\eta;h_o;\kappa]\right| < \infty \ \text{ for } \mathrm{n}=0,\ \dots,\ \mathrm{n_{max}} \tag{71}$$

and

$$ \lim_{\eta\to\pm\infty}\left|{}_i\rho^{-\frac{1}{2}}[\eta;\kappa]\,ld\,{}_i\psi_{\dagger,2\mathrm{j}}[\eta;h_o;\kappa]\right| < \infty \tag{72}$$

coupled with (52), we conclude that q-RSs (70) satisfy DBCs (62) and therefore represent an eigenfunction of SLE (61) with the eigenvalues ${}_i\varepsilon_{\mathbf{c},n}(h_o;\kappa)$. We thus proved that the RLDT in question keeps unchanged all $\mathrm{n_{max}}$ + 1 eigenvalues of prime SLE (49):

$$ {}_i\varepsilon_{\mathbf{c},\tilde{n}_n}(h_o;\kappa|\dagger,2\mathrm{j}) = {}_i\varepsilon_{\mathbf{c},n}(h_o;\kappa) \qquad \text{for } \mathrm{n}=0,\ \dots,\ \mathrm{n_{max}}\ \left(\dagger = \mathbf{d} \text{ or } \mathbf{d}'\right). \tag{73}$$

If prime SLE (61) has no additional eigenvalues between ${}_i\varepsilon_{\dagger,m}(h_o;\kappa)$ and ${}_i\varepsilon_{\mathbf{c},n_{\max}}(h_o;\kappa)$, then

$$ \tilde{n}_n = n+1 \tag{74}$$

and

$$ {}_i\Psi_{\mathbf{c},n+1}[\eta;h_o;\kappa|\dagger,2\mathrm{j}] = \Psi\left[\eta;h_o;\kappa;{}_i\varepsilon_{\mathbf{c},n}(h_o;\kappa)|\dagger,2\mathrm{j}\right]. \tag{75}$$

Let us now prove that eigenvalues (73) cover all the energy spectrum above the lowest-energy level ${}_i\varepsilon_{\dagger,2\mathrm{j}}(h_o;\kappa)$ or, in other words, that the given Dirichlet problem is exactly solvable. Indeed suppose that prime SLE (61) has another eigenfunction

$\tilde{\Psi}_{\mathbf{c},\tilde{n}}[\eta; h_o; \kappa|\dagger, 2\mathrm{j}]$ at an energy

$$\tilde{\varepsilon}(h_o; \kappa|\dagger, 2\mathrm{j}) \neq {}_i\varepsilon_{\mathbf{c},n}(h_o; \kappa|\dagger, 2\mathrm{j})\,for\ any\ n \geq 0, \tag{76}$$

so

$$\lim_{\eta\to\pm\infty} \tilde{\Psi}_{\mathbf{c},\tilde{n}}[\eta; h_o; \kappa|\dagger, 2\mathrm{j}] = 0. \tag{77}$$

Since solution (64) is nodeless, the eigenfunction in question must have at least one node.

Taking into account that RCSLE (60) has the second-order pole at infinity, we also affirm that this eigenfunction must obey the asymptotic formula

$$\lim_{\eta\to\pm\infty} \left|\eta\, ld\, \tilde{\Psi}_{\mathbf{c},\tilde{n}}[\eta; h_o; \kappa|\dagger, 2\mathrm{j}]\right| < \infty, \tag{78}$$

similar to (71) and (72).

Combining (78) with a similar limit for lowest-energy eigenfunction (64):

$$\lim_{\eta\to\pm\infty} \left|\eta\, ld\, {}_i\Psi_{\mathbf{c}0}[\eta; h_o; \kappa|\dagger, 2\mathrm{j}]\right| < \infty \tag{79}$$

and making use of (77) we conclude that the function

$$\begin{aligned}\tilde{\Psi}[\eta; h_o; \kappa] \equiv {}_i\rho^{-\frac{1}{2}}[\eta; \kappa]\, \tilde{\Psi}_{\mathbf{c},\tilde{n}}[\eta; h_o; \kappa|\dagger, 2\mathrm{j}] \\ \times\left\{ld\, \tilde{\Psi}_{\mathbf{c},\tilde{n}}[\eta; h_o; \kappa|\dagger, 2\mathrm{j}] - ld\, {}_i\psi_{\dagger,2\mathrm{j}}[\eta; h_o; \kappa]\right\}\end{aligned} \tag{80}$$

satisfies the DBCs at $\eta = \pm\infty$, and therefore, it must coincide with one of eigenfunctions of SLE (49), in contradiction with the assumption that eigenvalue (76) differs from any eigenvalue of the original Sturm-Liouville problem.

Setting

$${}_i\psi_{\dagger,\mathrm{m}}[\eta; h_o; \kappa] = \psi\left[\eta; {}_i\lambda_{\dagger,\mathrm{m}}\right] \mathfrak{R}_m^{\left({}_i\lambda_{\dagger,\mathrm{m}}\right)}[\eta], \tag{81}$$

where

$$\psi[\eta; \lambda] \equiv \left(\eta^2 + 1\right)^{-1/4} \phi[\eta; \lambda] = \sqrt[4]{\eta^2 + 1}\,(1 - i\eta)^{1/2\lambda}(1 + i\eta)^{1/2\lambda*} \tag{82}$$

one can directly verify that eigenfunctions (75) have a quasi-rational form

$${}_i\Psi_{\mathbf{c},n+1}[\eta; h_o; \kappa|\dagger, 2\mathrm{j}] = {}_i\psi_{\left[\eta; {}_i\lambda_{\mathbf{c},n}\right]} \frac{{}_i\mathrm{D}_{n+2\mathrm{j}+1}[\eta; h_o; \kappa|\dagger, 2\mathrm{j}; n]}{\sqrt{T_2[\eta; \kappa]}\,\mathfrak{R}_{2\mathrm{j}}^{\left({}_i\lambda_{\dagger,2\mathrm{j}}\right)}(\eta)}, \tag{83}$$

where the polynomial numerator of the fraction on the right represents the Routh-seed ($\mathcal{R}S$) "polynomial determinant" (PD)

$${}_i\mathrm{D}_{n+2\mathrm{j}+1}[\eta; h_o; \kappa|\dagger, 2\mathrm{j}; n] = \begin{vmatrix} \mathfrak{R}_m^{\left({}_i\lambda_{\dagger,2\mathrm{j}}\right)}[\eta] & \mathfrak{R}_n^{\left({}_i\lambda_{\mathbf{c},n}\right)}[\eta] \\ {}_iS_{m+1}^{\left({}_i\lambda_{\dagger,2\mathrm{j}}\right)}[\eta] & {}_iS_{n+1}^{\left({}_i\lambda_{\mathbf{c},n}\right)}[\eta] \end{vmatrix}, \tag{84}$$

with ${}_{i}S^{(\lambda)}_{m+1}[\eta]$ standing for the "$\mathcal{R}$S polynomial supplement" [49].

$$ {}_{i}S^{(\lambda)}_{m+1}[\eta] \equiv (\eta^2+1)\psi^{-1}[\eta;\lambda]\frac{d}{d\eta}\left(\psi^{-1}[\eta;\lambda]\mathfrak{R}^{(\lambda)}_{m}[\eta]\right) \tag{85} $$

$$ = [(Re\,\lambda+1)\eta + Im\,\lambda]\,\mathfrak{R}^{(\lambda)}_{m}[\eta] + (\eta^2+1)\dot{\mathfrak{R}}^{(\lambda)}_{m}[\eta]. \tag{86} $$

It will be demonstrated in next section that PDs (84) satisfy the Heine-type differential equations with degree-dependent linear coefficient functions so we refer to its polynomial solutions as "$\mathcal{R}$S Heine polynomials" despite the fact that they belong to different sets of (n + 1)(n + 2)/2 Heine polynomials depending on the choice of n.

5. $\mathcal{R}$S Heine polynomials

The gauge transformation

$$ {}_{i}\Phi[\eta;h_o;\kappa;\varepsilon|\dagger,2\mathrm{j}] = \phi\left[\eta;\kappa;{}_{i}\lambda_{\dagger,2\mathrm{j}};\lambda(h_o;\kappa;\varepsilon)\right] \times F[\eta;h_o;\kappa;\varepsilon|\dagger,2\mathrm{j}], \tag{87} $$

where

$$ \lambda(h_o;\kappa;\varepsilon) \equiv \sqrt{h_o+1+4(\kappa-1)\varepsilon} \tag{88} $$

is the root with the *positive* real part

$$ Re\,\lambda(h_o;\kappa;\varepsilon) > 0 \tag{89} $$

and

$$ \phi\left[\eta;\kappa;{}_{i}\lambda_{\dagger,2\mathrm{j}};\lambda\right] = \frac{\phi[\eta;-\lambda]}{\sqrt{T_2[\eta;\kappa]}\,\mathfrak{R}_{2\mathrm{j}}^{({}_{i}\lambda_{\dagger,2\mathrm{j}})}(\eta)}, \tag{90} $$

with the numerator defined via (82), converts RCSLE (60) into the ODE

$$ \mathbf{D}_{\dagger,2\mathrm{j}}[\eta;h_o;\kappa;\varepsilon]\,F[\eta;h_o;\kappa;\varepsilon|\dagger,2\mathrm{j}] = 0, \tag{91} $$

where

$$ \mathbf{D}_{\dagger,2\mathrm{j}}[\eta;h_o;\kappa;\varepsilon] \equiv A_{2\mathrm{j}+4}[\eta;h_o;\kappa|\dagger,2\mathrm{j}]\frac{d^2}{d\eta^2} \tag{92} $$

$$ +2B_{2\mathrm{j}+3}[\eta;h_o;\kappa;\varepsilon|\dagger,2\mathrm{j}]\frac{d}{d\eta} + C_{2\mathrm{j}+2}[\eta;h_o;\kappa;\varepsilon|\dagger,2\mathrm{j}] $$

is the second-order differential operator with polynomial coefficient functions, namely

$$ A_{2\mathrm{j}+4}[\eta;h_o;\kappa|\dagger,2\mathrm{j}] \equiv T_2[\eta;\kappa]\,(\eta^2+1)\,\mathfrak{R}_{2\mathrm{j}}^{({}_{i}\lambda_{\dagger,2\mathrm{j}})}[\eta], \tag{93} $$

and

$$B_{2j+3}[\eta; h_o; \kappa; \varepsilon|\dagger, 2j] \equiv T_2[\eta; \kappa](\eta^2+1)\,\mathfrak{R}_{2j}^{({}_i\lambda_{\dagger,2j})}[\eta] \tag{94}$$
$$\times\left(\frac{1-\lambda(h_o;\kappa;\varepsilon)}{2(\eta-i)} + c.c. - \sum_{l=1}^{2j}\frac{1}{\eta-\eta_{2j;\,l}({}_i\lambda_{\dagger,2j})} - \frac{\eta}{T_2[\eta;\kappa]}\right)$$

with $\eta_{2j;\,l}(\lambda)$ standing for 2j zeros of the Routh polynomial $\mathfrak{R}_{2j}^{(\lambda)}[\eta]$. The explicit form of the free-term of ODE (91) is nonessential in the current context, and we simply refer the reader to [3] for the specifics. The only important detail is that the polynomial coefficient function (94) is energy-dependent, and as a result, the polynomial coefficient function of the first derivative in Heine-type ODE (95) below depends on the polynomial degree.

Making solution (87) at $\varepsilon = {}_i\varepsilon_{\mathbf{c},n}(h_o;\kappa)$ coincide with eigenfunction (75) shows that PD (84) satisfies the Heine-type ODEs

$$\mathbf{D}_{\dagger;\,2j}\left[\eta; h_o; \kappa; {}_i\varepsilon_{\mathbf{c},n}(h_o;\kappa)\right]{}_i\mathrm{D}_{n+2j+1}[\eta; h_o; \kappa|\dagger, 2j; n] = 0. \tag{95}$$

The sequences of Gauss-seed Heine polynomials turn into finite or infinite sequences of EOPs if the ExpDiffs for poles in the finite plane are energy-independent [43], and as a result, the corresponding Liouville potential belongs to Group A of translationally shape-invariant (TSI) potentials in Odake-Sasaki's [44, 45] classification scheme. One of such "exceptional" cases will be discussed in next section.

Note that

$$\lambda\left(h_o; \kappa; {}_i\varepsilon_{\mathbf{d},2j}\right) = {}_i\lambda_{\mathbf{d},2j} > 0 \tag{96}$$

which implies that the free term of ODE (91) with $\dagger = \mathbf{d}$ vanishes at $\varepsilon = {}_i\varepsilon_{\mathbf{d},2j}$, and therefore, the ODE has a constant solution at this energy. The polynomial sequence specifying eigenfunctions of prime SLE (61) solved under the DBCs thus starts from a constant but lacks 2j polynomials of higher degrees. It will we proven in Section 7 that these polynomials turn into EOPs at κ = 1.

6. Translational form-invariance of RRef CSLE with simple pole density function

In the limit $\kappa \to 1$, the density function

$${}_i\rho_\diamond[\eta] \equiv {}_i\rho[\eta; 1] = \frac{1}{\eta^2+1} \tag{97}$$

has simple poles at $\pm i$. As a result of such a very specific choice of the density function the corresponding CSLE

$$\left\{\frac{d^2}{d\eta^2} + {}_iI^o[\eta; h_o] + \varepsilon\,{}_i\rho_\diamond[\eta]\right\}{}_i\Phi[\eta; h_o; 1; \varepsilon] = 0 \tag{98}$$

becomes TFI [46], that is, it has two basic solutions

$$ {}_i\phi_{\pm,0}[\eta;\lambda_o] = {}_i\rho_\diamond^{-\frac{1}{2}}[\eta]\,(1+i\eta)^{\pm\frac{1}{2}\lambda_o}(1-i\eta)^{\pm½\lambda_o^*} \tag{99}$$

which satisfy the generic translational form-invariance condition

$$ {}_i\phi_{-,0}[\eta;a+1+ib]_i\phi_{+;0}[\eta;a+ib] = {}_i\rho_\diamond^{-\frac{1}{2}}[\eta] \equiv {}_i\sigma^{\frac{1}{2}}[\eta], \tag{100}$$

where we use labels "+" and "-" instead of **d** and **d**′ (or **c**) accordingly and also define the TFI parameters a and b as the real and imaginary parts of the square root

$$ \lambda_o \equiv a+ib = \sqrt{h_o+1}\,(Re\,\lambda_o > 0). \tag{101}$$

It is remarkable that the cubic term in quartic eq. (41) disappears so the latter equation turns into the quadratic equation

$$ \lambda^4_{\pm,m;R} - \left(a^2-b^2\right)\lambda^2_{\pm,m;R} - a^2b^2 = 0 \tag{102}$$

with respect to $\lambda^2_{\pm,m;R}$ [21]. The crucial point is that the coefficients of this quadratic equation are independent of m, and as a result, the indexes of the Routh polynomials in the right-hand side of (42) and (43) become independent of the polynomial degree m. As originally pointed to by the author [43], the energy independence of the ExpDiffs for the poles in the finite complex plane is the necessary and sufficient condition for the RLDTs of the given RCSLE to be quantized in terms of EOPs. Concurrently with [43], Odake and Sasaki [44, 45] grouped all the TSI potentials of this kind into the so-called "Group A" associated with TFI SLEs with energy-independent ExpDiffs for the poles in the finite complex plane. The same year, Quesne [24] scrupulously studied rational SUSY partners of the Scarf II potential (the Gendenshtein potential [52, 53] in our terms) speculating that there exist nodeless FFs of type **d**, which can be thereby used to construct finite sequences of EOPs formed by RDTs of R-Routh polynomials (referred to by her simply as "Romanovski polynomials"). Her conjecture stimulated our intense interest in this issue [21].

Keeping in mind that the sought-for root ${}_i\lambda^2_{\pm,\mathrm{m};R}$ of the quadratic equation must be positive, one finds

$$ {}_i\lambda^2_{\pm,\mathrm{m};R} = a^2, \tag{103}$$

that is,

$$ {}_i\lambda_{\pm,\mathrm{m};R} = \pm a, \tag{104}$$

where we use labels "+" and "-" instead of **d** and **d**'(or **c**) accordingly. Substituting. (104) into (47) thus gives

$$ {}_i\rho_{\infty;\pm,m} = \mp a - m - 1 \tag{105}$$

so (37) takes the elementary form:

$$ {}_i\varepsilon_{\pm,m}(a) = -\left(\pm a + m + ½\right)^2 \tag{106}$$

Since the ExpDiffs for the poles of CSLE (98) in the finite complex plane are energy-independent, it is the CSLE of Group A [45, 46], and as a result, ChExps of q-RSs (42) and (43) for these poles become independent of the polynomial degree:

$$ {}_{i}\phi_{\pm,\mathrm{m}}[\eta;\lambda_o] = {}_{i}\mathrm{C}_{\pm,m}\,{}_{i}\phi_{\pm,0}[\eta;\lambda_o]\,\mathfrak{R}_m^{(\pm\lambda_o)}[\eta], \tag{107} $$

where the constant factors [3].

$$ {}_{i}C_m = 2^m m! \tag{108} $$

are selected in such a way that q-RSs (107) satisfy the generic ladder relations

$$ {}_{i}w[\eta;a,b|\mp,0;\ \pm,m] = {}_{i}\phi_{\pm,m+1}[\eta;a{\mp}1,b]/{}_{i}\phi_{\pm,0}[\eta;a{\mp}1,b] \tag{109} $$

derived by us [46] for TFI SLEs, with

$$ {}_{i}w[\eta;a,b|\pm,m;\mp,m'] \equiv W\{{}_{i}\phi_{\pm,m}[\eta;a,b],{}_{i}\phi_{\mp,m'}[\eta;a,b]\} \tag{110} $$

The sequences start from basic solutions (99) respectively.
The gauge transformations

$$ {}_{i}\Phi[\eta;a+ib;\varepsilon] \equiv {}_{i}\Phi\left[\eta;(a+ib)^2-1;1;\varepsilon\right] = $$

$$ {}_{i}\phi_{\pm;\,0}[\eta;a+ib]\,{}_{i}F_{\pm}[\eta;a+ib;\varepsilon] \tag{111} $$

convert CSLE (98) into a pair of Bochner-type eigenequations

$$ \left\{(\eta^2+1)\frac{d^2}{d\eta^2} + {}_{i}\tau_{\pm}[\eta;a,b]\frac{d}{d\eta} + \varepsilon - {}_{i}\varepsilon_{\pm,0}(a)\right\}{}_{i}F_{\pm}[\eta;a+ib;\varepsilon] = 0, \tag{112} $$

where

$$ {}_{i}\tau_{\pm}[\eta;a,b] \equiv 2ld_{i}\phi_{\pm,0}[\eta;a+ib] = 2(1\pm a)\eta \pm 2b. \tag{113} $$

Taking into account that

$$ {}_{i}\varepsilon_{\pm,m}(a) - {}_{i}\varepsilon_{\pm,0}(a) = m\,(1\pm 2a) \tag{114} $$

and comparing (112) and (113) with (9.9.5) in [14], we find that eigenequations (112) have polynomial solutions at energies (106):

$$ \left\{(\eta^2+1)\frac{d^2}{d\eta^2} + {}_{i}\tau_{\pm}[\eta;a,b]\frac{d}{d\eta} + m\,(1\pm 2a)\right\}P_n(x;b,\mp a-1) = 0. \tag{115} $$

The basic solution (99) is thus nothing but constant solution of these eigenequations converted back via gauge transformation (111). It is also worth mentioning that the solutions ${}_{i}\mathrm{R}[\eta;\mp a,\pm b]$ of Routh ODE (6) written as

$$ ld_{i}\mathrm{R}^{\pm}[\eta;a,b] = {}_{i}\tau_{\pm}[\eta;a,b]/(\eta^2+1) \tag{116} $$

coincide with the squares of basic solutions (99).

It is crucial all q-RSs (107) with the upper index lie below the lowest-energy level:

$$ {}_i\varepsilon_{+,m}(a) = -\left(a + m + ½\right)^2 < {}_i\varepsilon_{-,0}(a) = -\left(a - ½\right)^2 \tag{117} $$

and therefore, according to the disconjugacy theorem, any Routh polynomial $\mathfrak{R}^{(a+ib)}_{\mathrm{m}}[\eta]$ of an even (odd) degree m does not have real roots (has one and only one real root) if the real part of its index is positive. The latter assertion also holds for Routh polynomials $\mathfrak{R}^{(-a-ib)}_m[\eta]$ provided that

$$ {}_i\varepsilon_{-,m}(a) = -\left(m + ½ - a\right)^2 < {}_i\varepsilon_{-,0}(a), \tag{118} $$

that is, if $m > 2a - 1 > 0$.

7. Quantization of rational Darboux transforms of Gendenshtein potential by finite sequences of EOPs

The RLDT with the nodeless FF ${}_i\phi_{\pm,2\mathrm{j}}[\eta;\lambda_o]$ brings us to the RCSLE

$$ \left\{\frac{d^2}{d\eta^2} + {}_iI^o[\eta;\lambda_o|\pm,2\mathrm{j}] + \varepsilon\, {}_i\boldsymbol{\rho}_\diamond[\eta]\right\} {}_i\Phi[\eta;\lambda_o;\varepsilon|\pm,2\mathrm{j}] = 0. \tag{119} $$

Representing the given RefPF

$$ {}_iI^o[\eta;\lambda_o|\pm,2\mathrm{j}] = {}_iI^o[\eta;\lambda_o] + 2\sqrt{{}_i\boldsymbol{\rho}_\diamond[\eta]}\frac{d}{d\eta}\frac{ld\,{}_i\phi_{\pm,2\mathrm{j}}[\eta;\lambda_o]}{\sqrt{{}_i\boldsymbol{\rho}_\diamond[\eta]}} + \mathrm{J}\{{}_i\boldsymbol{\rho}_\diamond[\eta]\} \tag{120} $$

where the so-called [49] "universal correction" is defined via the generic formula

$$ \mathrm{J}\{\mathrm{f}[\eta]\} \equiv ½\sqrt{\mathrm{f}[\eta]}\frac{d}{d\eta}\frac{ld\,\mathrm{f}[\eta]}{\sqrt{\mathrm{f}[\eta]}}, \tag{121} $$

one finds [24].

$$ {}_iI^o[\eta;\lambda_o|\pm,2\mathrm{j}] = {}_iI^o[\eta;|\lambda_o \pm 1|] + 2\dot{ld}\,\mathfrak{R}^{(\pm\lambda_o)}_{2\mathrm{j}}[\eta] + \frac{2\eta}{\eta^2+1}ld\,\mathfrak{R}^{(\pm\lambda_o)}_{2\mathrm{j}}[\eta] \tag{122} $$

that is, the RLDTs in question change by ±1 the ExpDiffs for the poles at +*i* and -*i*. Let us set

$$ {}_i\mathrm{D}^{(-\lambda_o)}_{m+2\mathrm{j}+1}[\eta|+,2\mathrm{j};-,m] = \begin{vmatrix} \mathfrak{R}^{(\lambda_o)}_{2\mathrm{j}}[\eta] & \mathfrak{R}^{(-\lambda_o)}_{m}[\eta] \\ {}_i\mathrm{S}^{(\lambda_o)}_{2\mathrm{j}+1}[\eta] & {}_i\mathrm{S}^{(-\lambda_o)}_{m+1}[\eta] \end{vmatrix}, \tag{123} $$

where m is an arbitrary non-negative integer, and demonstrate that the polynomials in question, coupled with a constant, form an "exceptional" [54] DPS (X-DPS), which lacks 2j sequential polynomial degrees starting from 1 and thereby does not obey the prerequisites of the Bochner theorem [9]. To prove this assertion, first note that the power exponents of the q-RS

$$ {}_{i}\phi[\eta;\lambda_o|+,2\mathrm{j};\,-,m] = {}_{i}\phi_{\mathbf{c},0}[\eta;\lambda_o+1]\frac{{}_{i}\mathrm{D}^{(-\lambda_o)}_{m+2\mathrm{j}+1}[\eta|+,2\mathrm{j};\,-,m]}{\mathfrak{R}^{(\lambda_o)}_{2\mathrm{j}}(\eta)} \tag{124} $$

for the singular points $\pm i$ coincide with the ChExps for the respective poles of RCSLE (119) and therefore the numerator of the PF in the right-hand side of (124) may not have zeros at these singular points (at least as far as a is not a positive integer). Rewriting (124) as

$$ {}_{i}\phi[\eta;\lambda_o|+,2\mathrm{j};\,-,m] = {}_{i}\phi_{\mathbf{c},0}[\eta;\lambda_o|+,2\mathrm{j}]\,{}_{i}\mathrm{D}^{(-\lambda_o)}_{m+2\mathrm{j}+1}[\eta|+,2\mathrm{j};\,-,m], \tag{125} $$

and making gauge transformation (87) but now with the energy-independent gauge function

$$ {}_{i}\phi_{\mathbf{c},0}[\eta;\lambda_o|+,2\mathrm{j}] = \frac{{}_{i}\phi_{\mathbf{c},0}[\eta;\lambda_o+1]}{\mathfrak{R}^{(\lambda_o)}_{2\mathrm{j}}[\eta]} \tag{126} $$

representing the lowest-energy eigenfunction of RCSLE (119), we come to the Bochner-type equation with energy-independent polynomial coefficient functions of both first and second derivatives. As a direct corollary of this result, we conclude that PDs (123) satisfy the Bochner-type equation

$$ \mathbf{D}_{+;2\mathrm{j}}\left[\eta;\lambda_o;{}_{i}\varepsilon_{-,m}(\lambda_o)\right]{}_{i}\mathrm{D}^{(-\lambda_o)}_{m+2\mathrm{j}+1}[\eta|+,2\mathrm{j};\,-,m] = 0, \tag{127} $$

where

$$ \mathbf{D}_{+;2\mathrm{j}}[\eta;\lambda_o;\varepsilon] \equiv (\eta^2+1)\,\mathfrak{R}^{(\lambda_o)}_{2\mathrm{j}}[\eta]\frac{d^2}{d\eta^2} + {}_{i}\tau_{2\mathrm{j}+1}[\eta;\lambda_o|+,2\mathrm{j}]\frac{d}{d\eta} + C_{2\mathrm{j}}[\eta;\lambda_o;\varepsilon|+,2\mathrm{j}] \tag{128} $$

is the second-order differential operator with the coefficient function of the first derivative

$$ {}_{i}\tau_{2\mathrm{j}+1}[\eta;a+ib|+,2\mathrm{j}] = {}_{i}\tau_{-}[\eta;a+1,b]\,\mathfrak{R}^{(a+ib)}_{2\mathrm{j}}[\eta] \\ -2(\eta^2+1)\dot{\mathfrak{R}}^{(a+ib)}_{2\mathrm{j}}[\eta]. \tag{129} $$

The important new feature of differential operator (128), compared with (92), is that polynomial coefficient (129) is energy-independent and therefore the free term of differential operator (128) necessarily coincides with one of the “characteristic polynomials” [55] of the Heine equation under consideration while PD (123) represents the corresponding Heine polynomial. It directly follows from our choice (126) of the gauge function that

$$ C_{2\mathrm{j}}\left[\eta;\lambda_o;{}_{i}\varepsilon_{+,2\mathrm{j}}(\lambda_o)|+,2\mathrm{j}\right] = 0. \tag{130} $$

One can directly verify that the q-RSs of the associate prime equation

$$ {}_i\Psi_{\mathrm{c,n+1}}[\eta;\lambda_o|+,2\mathrm{j}] = {}_i\Psi_{\mathrm{c},0}[\eta;\lambda_o+1]\frac{{}_i\mathrm{D}^{(-2b,1-a)}_{n+2\mathrm{j}+1}[\eta|+,2\mathrm{j};-,n]}{\mathfrak{R}^{(\lambda_o)}_{2\mathrm{j}}(\eta)}, \tag{131} $$

where

$$ {}_i\mathrm{D}^{(-2b,1-a)}_{n+2\mathrm{j}+1}[\eta|+,2\mathrm{j};-,n] \equiv {}_i\mathrm{D}^{(-a-ib)}_{n+2\mathrm{j}+1}[\eta|+,2\mathrm{j};-,n] \tag{132} $$
$$ = \begin{vmatrix} \mathfrak{R}^{(a+ib)}_{2\mathrm{j}}[\eta] & R^{(-2b,1-a)}_n(\eta) \\ {}_i\mathrm{S}^{(a+ib)}_{2\mathrm{j}+1}[\eta] & {}_i\mathrm{S}^{(-a-ib)}_{n+1}[\eta] \end{vmatrix} \quad \text{for } 0 \leq n < a - ½, $$

vanish at infinity and thereby the RDT of the nth eigenfunction of RRef CSLE (98) is itself the eigenfunction of RCSLE (119) with the eigenvalue ${}_i\varepsilon_{\mathrm{c},n}(\lambda_o)$. Since the lowest-energy eigenfunction is represented by nodeless solution (126) with the eigenvalue ${}_i\varepsilon_{+,2\mathrm{j}}(\lambda_o)$, there is a jump in the degrees of polynomials forming a complete set of $\lceil a + ½ \rceil$ eigenfunctions of RCSLE (119).

Keeping in mind that eigenfunctions (131) are orthogonal with the weight

$$ {}_iw[\eta] = (\eta^2+1)^{-½} \tag{133} $$

PDs (132) must be orthogonal with the weight

$$ \mathrm{W}[\eta;\lambda_o|+,2\mathrm{j}] \equiv \frac{{}_i\Psi^2_{\mathrm{c},0}[\eta;\lambda_o+1]}{\left|\mathfrak{R}^{(-\lambda_o)}_{2\mathrm{j}}(\eta)\right|^2}\,{}_iw[\eta]. \tag{134} $$

All PDs (132) also obey the linear orthogonality condition:

$$ \int\limits_{-\infty}^{-\infty} {}_i\mathrm{D}^{(-2b,1-a)}_{n+2\mathrm{j}+1}[\eta|+,2\mathrm{j};-,n]\,\mathrm{W}[\eta;\lambda_o|+,2\mathrm{j}]d\eta = 0 \tag{135} $$

for $0 \leq \mathrm{n} \leq \lfloor a - ½ \rfloor$.

We thus deal with a finite sequence of EOPs, which starts from a constant but lacks 2j polynomials of higher degree starting from the first-degree polynomial so the prerequisites of the Bochner theorem [9] do not hold.

Coming back to RCSLE (119), first note that its q-RSs

$$ {}_i\phi[\eta;\lambda_o|-,2\mathrm{j};\mathrm{n}] = \frac{W\left\{{}_i\phi_{-,2\mathrm{j}}[\eta;\lambda_o],{}_i\phi_{\mathrm{c},n}[\eta;\lambda_o]\right\}}{{}_i\phi_{-,2\mathrm{j}}[\eta;\lambda_o]}, \tag{136} $$

where n is an arbitrary non-negative integer, can be expressed in terms of the Wronskians of two Routh polynomials:

$$ {}_i\mathrm{W}^{(-\lambda_o)}_{2\mathrm{j}+n-1}[\eta|2\mathrm{j},n] \equiv W\left\{\mathfrak{R}^{(-\lambda_o)}_{2\mathrm{j}}[\eta],\mathfrak{R}^{(-\lambda_o)}_{n}[\eta]\right\} \tag{137} $$

as follows

$$ {}_i\phi[\eta;\lambda_o|-,2\mathrm{j};\mathrm{n}] = -g[\eta;\lambda_o|2\mathrm{j}]\,{}_i\mathrm{W}^{(-\lambda_o)}_{2\mathrm{j}+n-1}[\eta|n,2\mathrm{j}], \tag{138} $$

where

$$g[\eta;\lambda_o|2j] \equiv \frac{{}_i\phi[\eta;\lambda_o-1]}{\mathfrak{R}_{2j}^{(-\lambda_o)}(\eta)}. \tag{139}$$

Since the power exponents of this solution for the singular points $\pm i$ coincide with the ChExps for the respective poles of RCSLE (119) polynomial Wronskian (137) may not have zeros at these singular points (again as far as the parameter a is not a positive integer). As a direct corollary of this assertion, we conclude that the polynomial Wronskians in question must satisfy the Bochner-type differential equation with the polynomial coefficient of the first derivative independent of the polynomial degree and thereby form an X-DPS starting from the polynomial of degree 2j − 1. In contrast with the X-DPS formed by PDs (123), the X-DPS composed of the polynomial Wronskians does not contain a polynomial of zero degree.

One can verify that the quasi-rational solution of the corresponding prime SLE,

$${}_i\psi[\eta;\lambda_o|-,2j;n] = -(\eta^2+1)^{-\frac{1}{4}} g[\eta;\lambda_o|2j]\,{}_i\mathrm{W}_{2j+n-1}^{(-\lambda_o)}[\eta|n,2j], \tag{140}$$

vanishes at infinity iff $n \le \lfloor a - 1/2 \rfloor$ and therefore represents the (n + 1)-th eigenfunction

$${}_i\psi_{c,n+1}[\eta;a+ib|-,2j] = -(\eta^2+1)^{-\frac{1}{4}} g[\eta;a+ib|2j]\,{}_i\mathrm{W}_{n+2j-1}^{(-2b,1-a)}[\eta] \tag{141}$$

where

$${}_i\mathrm{W}_{n+2j-1}^{(-2b,1-a)}[\eta] \equiv W\left\{\mathfrak{R}_{2j}^{(-\lambda_o)}[\eta], \mathrm{R}_n^{(-2b,1-a)}(\eta)\right\}. \tag{142}$$

Since eigenfunctions (141) are orthogonal with weight (133) polynomial Wronskians (142) must be orthogonal with the weight

$$\mathrm{W}[\eta;a+ib|-,2j] \equiv g^2[\eta;a+ib|2j]/(\eta^2+1). \tag{143}$$

We thus deal with a finite sequence of EOPs, which starts from a polynomial of degree $2j - 1 \ge 1$

$${}_i\mathrm{W}_{2j-1}^{(-2b,1-a)}[\eta] = -\dot{\mathfrak{R}}_{2j}^{(-\lambda_o)}[\eta] \tag{144}$$

so the prerequisites of the Bochner theorem [9] do not hold. Since lowest-energy eigenfunction

$${}_i\phi_{c,0}[\eta;\lambda_o|-,2j] = \frac{(1-i\eta)^{\frac{1}{2}\lambda_o}(1+i\eta)^{\frac{1}{2}\lambda_o^*}}{\mathfrak{R}_{2j}^{(-\lambda_o)}[\eta]} \tag{145}$$

represents a solution of RCSLE (119) with the opposite ChExp (compared with higher-energy eigenfunctions), the given sequence of EOPs does not start from a constant in contrast with the finite EOP sequence composed of PDs.

8. Conclusions

The presented analysis points to the important distinction between Routh polynomials [1] and its finite orthogonal subset revealed by Romanovsky [17] a few decades later (with no connection to Routh' paper ignored by mathematicians for more than a century). We refer to the latter subset as "Romanovski-Routh" polynomials (similarly to the terms "Romanovski-Jacobi" and "Romanovski-Bessel" polynomials suggested by Lesky [18, 19] for two other finite sequences of the orthogonal polynomials discovered by Romanovsky). It was shown that there are two infinite sequences of Routh polynomials without real roots (in contrast with R-Routh polynomials with all the roots located on the real axis). The FFs formed by polynomials from these sequences can be thus used to generate new rational Sturm-Liouville equations exactly solvable under the DBCs. This observation opens a novel area of applications for the Routh DPS—a little known real-field reduction of the complex Jacobi DPS left without proper attention for decades [12] in the shadow of its powerful sibling composed of real Jacobi polynomials.

The current analysis is focused solely on RRef CSLE (19) with the density function having two simple zeros on the opposite sides of the imaginary axis [20, 21]. (We refer the reader to [3] for the general case of the density function with two complex conjugated zeros away from the imaginary axis.) A special attention was given to the limiting case of the density function with the simple poles at $\pm i$ when the RRef CSLE becomes TFI. Based on the disconjugacy theorem [33–37], it was proved that each Routh polynomial $\mathfrak{R}_{2j}^{(a+ib)}[\eta]$ of even degree with $a > 0$ may not have real roots. Similarly all other Routh polynomials from this sequence may have only one (necessarily simple) real root. Any of the mentioned Routh polynomials of even degree can be used for constructing a finite EOP sequence [24], which starts from a constant and lacks as minimum two polynomials of the first and second degrees.

There also exists the second infinite family of the RDTs of R-Routh polynomials, which is composed of finite EOP sequences of Wronskians of Routh and R-Routh polynomials.

As originally proven in [46] and scrupulously examined in [56], one can construct the complete net of finite EOP sequences formed by rational Darboux-Crum transforms (RDCTs) of R-Routh polynomials using only the FFs composed of Wronskians of Routh polynomials $\mathfrak{R}_{2j}^{(-a-ib)}[\eta]$ and "juxtaposed" [57–59] pairs of R-Routh polynomials. The RDTs of R-Routh polynomials using the FFs $_i\phi_{-,2j}[\eta; \lambda_o]$ represent the simplest example of the finite EOP sequences constructed in such a way. On other hand, as discussed in detail in [3], the infinite sequence of RCSLEs obtained by means of the FFs $_i\phi_{+,2j}[\eta; \lambda_o]$ can be alternatively constructed using 2j seed functions $_i\phi_{-,n}[\eta; \lambda_o]$ composed of R-Routh polynomials of degrees n = 1, … , 2j.

Acknowledgements

I am grateful to Mariana Kirchbach for numerous thoughtful comments on Routh's paper, which notably shaped my current perception of the subject. Christiane Quesne's conjecture on existence of Routh polynomials without real zeros played a crucial role in this research. I am especially thankful to Mourad Ismail for bringing my attention to quasi-definite orthogonality of Routh polynomials, which helped to rectify my terminology.

Author details

Gregory Natanson†
a.i. solutions Inc., Lanham, Maryland, U.S.A

*Address all correspondence to: gregorynatanson@gmail.com

†Retired.

References

[1] Routh EJ. On some properties of certain solutions of a differential equation of second order. Proceedings of the London Mathematical Society. 1884; **16**:245-261. DOI: 10.1112/plms/s1-16.1.245

[2] Ismail MEH. Classical and Quantum Orthogonal Polynomials in One Variable. Cambridge: University Press; 2005. pp. 508-509. DOI: 10.1007/BF02985757

[3] Natanson G. Routh polynomials: hundred years in obscurity. 2022. pp. 1-61. Available from: http://www.researchgate.net/publication/326522529. DOI: 10.13140/RG.2.2.18824.32008/2

[4] Rodrigues O. De l'attraction des sphéroïdes. Correspondence sur l'École Impériale Polytechnique. 1816;**3**:361, 385 ISBN: 2019220997

[5] Jacobi GJ. Ueber eine besondere Gattung algebraischer Functionen, die aus der Entwicklung der Function $(1 - 2xz + z^2)^{1/2}$ entstehen. Journal fur die Reine und Angewandte Mathematik. 1827;**2**:223-226. DOI: 10.1515/crll.1827.2.223

[6] Hildebrandt EH. Systems of polynomials connected with the Charlier expansions and the Pearson differential and difference equations. Annals of Mathematical Statistics. 1931;**2**:379-439. http://projecteuclid.org/euclid.aoms/1177732950. DOI: 10.1214/aoms/1177732950

[7] Pearson K. Contributions to the mathematical theory of evolution, II: Skew variation in homogeneous material. Philosophical Transactions of the Royal Society A. 1895;**186**:343-424 http://bayes.wustl.edu/Manual/Pearson_1895.pdf. DOI: 10.1098/rsta.1895.0010

[8] Nikiforov AF, Uvarov VB. Special Functions of Mathematical Physics. Basel: Birkhåuser Verlag; 1988. DOI: 10.1007/978-1-4757-1595-8

[9] Bochner S. Über Sturm-Liouvillesche Polynomsysteme. Mathematische Zeitschrift. 1929;**29**:730-736 http://eudml.org/doc/168099. DOI: 10.1007/BF01180560

[10] Everitt WN, Littlejohn LL. Orthogonal polynomials and spectral theory: A survey. In: Brezinski C, Gori RA, editors. Orthogonal Polynomials and their Applications. IMACS Annals on Computing and Applied Mathematics. Vol. 9. Basel: J.C. Baltzer; 1991. pp. 21-55 ISSN 1012-2435

[11] Everitt WN, Kwon KH, Littlejohn LL, Wellman R. Orthogonal polynomial solutions of linear ordinary differential equations. Journal of Computationsl & Applied Mathematics. 2001;**133**:85-109. http://www.sciencedirect.com/science/article/pii/S0377042700006361.DOI:. DOI: 10.1016/S0377-0427(00)00636-1

[12] Kwon KH, Littlejohn LL. Classification of classical orthogonal polynomials. Journal of the Korean Mathematical Society. 1997;**34**:973-1008 https://www.researchgate.net/publication/238203450. Corpus ID:123032920

[13] Cryer CW. Rodrigues' formulas and the classical orthogonal polynomials. Bolletino dell Unione Matematica Italiana. 1970;**25**:1-11

[14] Koekoek R, Lesky PA, Swarttouw RF. Hypergeometric Orthogonal Polynomials and their Q-Analogues. Heidelberg: Springer; 2010. DOI: 10.1007/978-3-642-05014-5

[15] Koornwinder T. Additions to the formula lists in 'Hypergeometric orthogonal polynomials and their q-analogues' by Koekoek. Lesky and Swarttouw. 2022. pp. 1-45. Available from http://staff.fnwi.uva.nl/t.h.koornwinder/art/informal/KLSadd.pdf

[16] Jordaan K, Toókos F. Orthogonality and asymptotics of pseudo-Jacobi polynomials for non-classical parameters. Journal of Approximation Theory. 2014; **178**:1-12. http://sciencedirect.com/science/article/pii/S0021904513001664. DOI: 10.1016/j.jat.2013.10.003

[17] Romanovski VI. Sur quelques classes nouvelles de polynomes orthogonaux. Comptes Rendus. Académie des Sciences. 1929;**188**:1023-1025

[18] Lesky PA. Vervollstandigungderklassischen Orthogonalpolynomedurch Erganzungen zum Askey - Schema der hypergeometrischen orthogonalen Polynome. Ost Ak Wiss. 1995;**204**:151-166

[19] Lesky PA. Endliche und unendliche Systeme von kontinuierlichen klassichen Othogonalpolynomen. Zeitschrift für Angewandte Mathematik und Mechanik. 1996;**76**:181-184. DOI: 10.1002/zamm.19960760317

[20] Milson R. Liouville transformation and exactly solvable Schrödinger equations. International Journal of Theoretical Physics. 1998;**37**:1735-1752. arXiv.org/abs/solv-int/9706007. DOI: 0.1023/A:1026696709617

[21] Natanson G. Exact quantization of the Milson potential via Romanovski-Routh polynomials. 2015. pp.1-61. Available from: arXiv:1310.0796v3. DOI: 10.13140/RG.2.2.24354.09928/1

[22] Bose AK. A class of solvable potentials. Nuovo Cimento. 1964;**32**: 679-688. DOI: 10.1007/BF02735890

[23] Compean CB, Kirchbach M. The trigonometric Rosen-Morse potential in the supersymmetric quantum mechanics and its exact solutions. Journal of Physics A. 2006;**39**:547-557. arXiv:quant-ph/0509055v2. DOI: 10.1088/0305-4470/39/3/007

[24] Quesne C. Extending Romanovski polynomials in quantum mechanics. Journal of Mathematical Physics. 2013; **54**(122103):15 arXiv:1308.2114v2. DOI: 0.1063/1.4835555

[25] Raposo AP, Weber HJ, Alvarez-Castillo DE, Kirchbach M. Romanovski polynomials in selected physics problems. Centr. European Journal of Physics. 2007;**5**:253-284. arXiv: 0706.3897v1. DOI: 10.2478/s11534-007-0018-5

[26] Avarez-Castillo DE, Kirchbach M. Exact spectrum and wave functions of the hyperbolic scarf potential in terms of finite Romanovski polynomials. Revista Mexicana de Fisica E. 2007;**53**:143-154 http://www.researchgate.net/publication/242322566

[27] Stevenson AF. Note on the 'Kepler problem' in a spherical space, and the factorization method of solving eigenvalue problems. Physics Review. 1941;**59**:842-843. DOI: 10.1103/PhysRev.59.842

[28] Natanson G. Quantization of One-Dimensional Ginocchio Potential by Masjed-Jamei Polynomials with Degree-Dependent Indexes. 2022. pp. 1-51. Available from: http://www.researchgate.net/publication/360950671

[29] Ginocchio J. A class of exactly solvable potentials: I. one-dimensional Schrödinger equation. Annalen der Physik. 1984;**152**:03-219. DOI: 10.1016/0003-4916(84)90084-8

[30] Natanzon GA. Use of the Darboux theorem for constructing the general solution of the Schrödinger equation with the Pöschl-teller potential. Vestnik Leningrad University. 1977;**16**:33-39 http://www.researchgate.net/publication/316150022 (in Russian)

[31] Andrianov AA, Borisov NV, Ioffe MV. The factorization method and quantum systems with equivalent energy spectra. Physics Letters A. 1984;**105**: 19-22. DOI: 10.1016/0375-9601(84)90553-X

[32] Sukumar CV. Supersymmetric quantum mechanics of one-dimensional system. Journal of Physics A. 1985;**18**: 2917-2936. DOI: 10.1088/0305-4470/18/15/020

[33] Grandati Y. New rational extensions of solvable potentials with finite bound state spectrum. Physics Letters A. 2012; **376**:2866-2872. arXiv:1203.4149v1. DOI: 10.1016/j.physleta.2012.09.037

[34] Grandati Y. Rational extensions of solvable potentials and exceptional orthogonal polynomials. Journal of Physics: Conference Series. 2012;**343**: 012041,11pp. http://www.researchgate.net/publication/241778882. DOI: 10.1088/1742-6596/343/1/012041

[35] Grandati Y, Bérard A. Comments on the generalized SUSY QM partnership for Darboux- Pöschl-teller potential and exceptional Jacobi polynomials. Journal of Engineering Mathematics. 2013;**82**: 161-171. DOI: 10.1007/s10665-012-9601-x

[36] Grandati Y, Quesne C. Disconjugacy, regularity of multi-indexed rationally-extended potentials, and Laguerre exceptional polynomials. Journal of Mathematical Physics. 2013;**54**: 073512,13pp. arXiv:1211.5308v2. DOI: 10.1063/1.4815997

[37] Gòmez-Ullate D, Grandati Y, Milson R. Extended Krein-Adler theorem for the translationally shape invariant potentials. Journal of Mathematical Physics. 2014;**55**:043510, 30pp. arXiv:1309.3756. DOI: 10.1063/1.4871443

[38] Natanson G. Supersymmetry of Algebraic-Fraction Potentials Analytically Quantized in Terms of Heine Polynomials II. Generalized Darboux Transformations with Basic Quasi-Algebraic Kernels. 2011. pp. 1-58. Available from: http://www.researchgate.net/publication/326678021. DOI: 10.13140/RG.2.2.16798.43846

[39] Cooper F, Ginocchio JN, Khare A. Relationship between supersymmetry and solvable potentials. Physical Review D. 1987;**36**:2458-2473. DOI: 10.1103/PhysRevD.36.2458

[40] Natanson G. Single-source nets of algebraically-quantized reflective Liouville potentials on the line I. Almost-everywhere holomorphic solutions of rational canonical Sturm-Liouville equations with second-order poles. 2015. pp. 1-112. Available from: arXiv: 1503.04798v2. DOI: 10.13140/RG.2.2.27895.70567

[41] Heine E. Handbuch der Kugelfunctionen, Theorie und Anwendungen. Vol. 1. Berlin: G. Reimer; 1878. pp. 172-173

[42] Szegö G. Orthogonal Polynomials. New York: American Mathematical Society; 1959. pp. 150-151 ISBN: 0821810235, 9780821810231

[43] Natanson G. Gauss-seed nets of Sturm-Liouville problems with energy independent characteristic exponents and related sequences of exceptional orthogonal polynomials I. Canonical Darboux transformations using almost

everywhere holomorphic factorization functions. 2013. pp. 1-89. Available from: arXiv:1305.7453v1. DOI: 10.48550/arXiv.1305.7453

[44] Odake S, Sasaki R. Extensions of solvable potentials with finitely many discrete eigenstates. Journal of Physics A. 2013;**46**:235205, 15 pp. arXiv: 1301.3980. DOI: 10.1088/1751-8113/46/23/235205

[45] Odake S, Sasaki R. Krein-Adler transformations for shape-invariant potentials and pseudo virtual states. Journal of Physics A. 2013;**46**:245201, 24pp. arXiv:1212.6595v2. DOI: 10.1088/1751-8113/46/24/245201

[46] Natanson G. Equivalence Relations for Darboux-Crum Transforms of Translationally form-Invariant Sturm-Liouville Equations. 2021. pp. 1-58. Available from: http://www.researchgate.net/publication/353131294. DOI: 10.13140/RG.2.2.36086.93768/1

[47] Quesne C. Revisiting (quasi-) exactly solvable rational extensions of the Morse potential. International Journal of Modern Physics A: Particles and Fields; Gravitation; Cosmology; Nuclear Physics. 2012;**27**:1250073, 18pp. arXiv:1203.1812v2. DOI: 10.1142/S0217751X1250073X

[48] Quesne C. Rational extensions of the Rosen-Morse II and Eckart potentials. SIGMA. 2012;**8**:080, 19pp. arXiv: 1208.6165v2. DOI: 10.3842/SIGMA.2012.080

[49] Natanson G. Darboux-Crum nets of Sturm-Liouville problems solvable by quasi-rational functions I. General Theory. 2018. pp. 1-99. Available from http://www.researchgate.net/publication/323831953. DOI: 10.13140/RG.2.2.31016.06405/1

[50] Gesztesy F, Simon B, Teschl G. Zeros of the Wronskian and renormalized oscillation theory. American Journal of Mathematics. 1996;**118**:571-594 math.caltech.edu/papers/bsimon/p252.pdf. DOI 10.1353/ajm.1996.0024

[51] Rudyak BV, Zakhariev BN. New exactly solvable models for Schrödinger equation. Inverse Problems. 1987;**3**: 125-133. DOI: 10.1088/0266-5611/3/1/014

[52] Gendenshtein LE. Derivation of exact spectra of the Schrödinger equation by means of supersymmetry. JETP Letters. 1983;**38**:356-359 jetpletters.ac.ru/ps/1822/article_27857.pdf

[53] Natanson G. On History of the Gendenshtein ('Scarf II') Potential. 2018. pp. 1-5. Available from: http://www.researchgate.net/publication/323831807. DOI: 10.13140/RG.2.2.20949.73444/1

[54] Gómez-Ullate D, Kamran N, Milson R. An extension of Bochner's problem: Exceptional invariant subspaces. Journal of Approximation Theory. 2010;**162**:987-1006. arXiv: 0805.3376v3. DOI: 10.1016/j.jat.2009.11.002

[55] Marden M. On Stieltjes polynomials. Transactions of the American Mathematical Society. 1931;**33**:934-944. DOI: 10.1090/S0002-9947-1931-1501624-1

[56] Natanson G. On Finite Exceptional Orthogonal Polynomial Sequences Composed of Wronskian Transforms of Romanovski-Routh Polynomials. 2022. pp. 1-38. Available from: http://www.researchgate.net/publication/364350392. DOI: 10.13140/RG.2.2.12788.83847

[57] Samsonov BF. On the equivalence of the integral and the differential exact solution generation methods for the one-

dimensional Schrodinger equation. Journal of Physics A. 1995;**28**:6989-6998. DOI: 10.1088/0305-4470/28/23/036

[58] Samsonov BF. New features in supersymmetry breakdown in quantum mechanics. Mod. Physics Letters A. 1996; **11**:1563-1567, arXiv:quant-ph/9611012v1. DOI: 10.1142/S0217732396001557

[59] Bagrov VG, Samsonov BF. Darboux transformation and elementary exact solutions of the Schrödinger equation. Pramana-Journal of Physics. 1997;**49**: 563-580, ias.ac.in/article/fulltext/pram/049/06/0563-0580. DOI: 10.1007/BF02848330

Chapter 6

The Inverse Characteristic Polynomial Problem for Graphs over Finite Fields

Charles R. Johnson, Greyson C. Wesley, Xiaoyu Lin, Xiwen Liu and Sihan Zhou

Abstract

Let $\mathbb{F}$ be a finite field, and let G be a graph on n vertices. We study the possible characteristic polynomials that may be realized by matrices A over a finite field such that the graph of A is G. We focus mainly on the case G is a tree T, not only because trees are computationally simpler, but also because the theory of eigenvalue multiplicities is much better understood for trees than it is for general graphs. We demonstrate the applications to this problem by branch duplication and the recently developed geometric Parter-Wiener, etc. theory. We end with a list of several conjectures which should pave the way for future study.

Keywords: polynomial, graph, matrix, field, multiplicity

1. Introduction

Let $\mathbb{F}$ be a field. We say that a combinatorially symmetric matrix $A = (a_{ij}) \in M_n(\mathbb{F})$ has (undirected, simple) graph G if $a_{ij} \neq 0$ precisely when $\{i,j\}$ is an edge of G (no restriction is placed upon the diagonal entries of A, except that they are in $\mathbb{F}$). We write $G(A) = G$ to indicate the graph of A is G. By $\mathcal{F}(G)$ we mean $\{A \in M_n(\mathbb{F}) : G(A) = G\}$. Each $A \in \mathcal{F}(G)$ has a monic characteristic polynomial of degree n with coefficients in $\mathbb{F}$. The **inverse characteristic polynomial problem** over $\mathbb{F}$ for G is to determine the set

$$P(\mathbb{F}, G) = \{p(x) = \det(xI - A) | A \in \mathcal{F}(G)\}. \tag{1}$$

If we denote the set of all monic, degree n polynomials over $\mathbb{F}$ by $P_n(\mathbb{F})$, we may ask which n-vertex graphs G satisfy $P(\mathbb{F}, G) = P_n(\mathbb{F})$; we call such G **constructible**. If a graph G is not constructible, we say G is **non-constructible**.

Our primary interest here is the case in which $\mathbb{F}$ is a finite field $\mathbb{F}_q$, the finite field of order q (where q is any prime power), in which case the number of monic degree n polynomials over $\mathbb{F}_q$, denoted $|P_n(\mathbb{F}_q)|$, is given by $|P_n(\mathbb{F}_q)| = |\mathbb{F}_q|^n = q^n$. Although we are interested in all graphs, we focus primarily on trees. We present a mixture of theoretical results and suggestive, extensive empirical/computational results.

In case $\mathbb{F} = \mathbb{C}$, the complex numbers, all graphs are constructible: given an n-by-n matrix A over $\mathbb{C}$ and complex numbers $\lambda_1, \ldots, \lambda_n \in \mathbb{C}$, then there exists a diagonal matrix D such that $A + D$ has eigenvalues $\lambda_1, \ldots, \lambda_n$ [1]. If $\mathbb{F} = \mathbb{R}$, the real numbers, the question has been studied [2], though a complete answer is not yet known. This already motivates our problem, and extensive recent work on multiplicities [3] does as well. In our case, many trees are non-constructible, and it is an engaging combinatorial problem to ask which trees are constructible over a given field.

The remaining content is organized as follows. The next section summarizes our empirical results, with some details in the Appendix. We hope these will be useful to other researchers, who may make further advances with these data as a guide. In the third section, we give some theoretical results, inspired partly by what we found. In the fourth section, we record a number of conjectures that appear strongly suggested by our data. Some of these seem plausible, and others are a bit of a surprise. As construction techniques are difficult to come by, these conjectures should inspire further work.

2. Empirical data

Given a tree T, let $N(T)$ be the subset of $\mathcal{F}(T)$ having all nonzero subdiagonal entries equal to 1. More generally, for any graph G, let $N(G)$ be the subset of $\mathcal{F}(G)$ such that for any $A \in N(G)$ there exists a spanning tree $T' \subset G$ whose nonzero subdiagonal entries correspond to edges in T' equal to 1. By diagonal similarity, the characteristic polynomial of any $A \in \mathcal{F}(G)$ is preserved if $n - 1$ entries are normalized (See, for example, [4]). It follows that $A \in \mathcal{F}(G)$ has $p_A(x) = f(x)$ if and only if there exists $A' \in N(G)$ such that $p_{A'}(x) = f(x)$. Therefore, to determine if G realizes $f(x)$, it suffices to compute $p_A(x)$ for all $A \in N(G)$.

Let $\mathbb{F} = \mathbb{F}_q$, the finite field of order q, where q is a prime power. Let G be a graph with n vertices and e edges. There are q choices for each of the n diagonal entries and $q - 1$ choices for each of the $2e - (n - 1)$ entries that correspond to the edges in $G - T'$. It follows that

$$|N(G)| = q^n (q-1)^{2e-n+1}. \tag{2}$$

This implies, for example, that if we want to show G is constructible, then it suffices to compute the characteristic polynomials of no more than $q^n (q-1)^{2e-n+1}$ matrices. If all polynomials are realized, then G is constructible; otherwise, G is non-constructible.

2.1 Acquisition of raw data

We wrote a program in Mathematica [5] to compute the characteristic polynomials of all matrices in $N(G)$, thereby determining if G is constructible over $\mathbb{F}_q$. By Equation

(2), Mathematica computes the characteristic polynomials of at most $q^n(q-1)^{2e-n+1}$ matrices for each n-vertex graph G on n vertices with e edges.

We acquired extensive data in the case the graph G is a tree T, and the field is the finite field $\mathbb{F} = \mathbb{F}_3$ since these conditions are computationally the lightest (after the exceptional case $\mathbb{F} = \mathbb{F}_2$). The realizability status for all polynomials f, including the number of matrices in $\mathcal{F}(T)$ realizing f, has been computed for all trees T up to $n = 10$ vertices. Because the program automatically normalizes $n-1$ entries, the number of times a given polynomial $f(x)$ is realized by G refers to the number of elements $A \in N(G)$ such that $p_A(x) = f(x)$.

The realizability status for each polynomial over $\mathbb{F}_3$ is also known for several trees T on $n = 11$ vertices, but data for the number of matrices in $N(T)$ that realize a given polynomial was not computed in exchange for program speed. This was done by allowing the program to forget the previous realizations of $f(x)$ by T during the computation and to only record whether or not $f(x)$ is realized by some $A \in N(T)$. If the program finds that G realizes all polynomials, then computation for that tree terminates, and this is recorded so that computation for a new tree can begin. (This is why most trees for which the constructibility status is known on 11 vertices are constructible.)

We then acquired data over the field $\mathbb{F} = \mathbb{F}_5$ for all trees through $n = 8$ vertices. However, again to promote speed, the program only records the realizability status of all polynomials and does not record the number of times each is realized.

Finally, we acquired data over the field $\mathbb{F} = \mathbb{F}_3$ for all connected graphs through $n = 7$ vertices. See the Appendix for instructions to access these data.

2.2 Graph of graphs

Definition 1 (co-realizable graphs). We say that graphs G and H are **co-realizable** (over $\mathbb{F}$) if $P(\mathbb{F}, G) = P(\mathbb{F}, H)$. We call the equivalence class of G with respect to co-realizability the **co-realizability class** of G (over $\mathbb{F}$).

We can equip the set of all co-realizability classes of graphs on n vertices with a natural partial order by declaring that $H < G$ if and only if $P(\mathbb{F}, H) \subset P(\mathbb{F}, G)$. Organizing the empirical data with respect to this partial order reveals the structure of co-realizable classes. An example demonstrating this graph in the case $\mathbb{F} = \mathbb{F}_3$, $n = 8$, together with additional information about each tree, can be found in **Figure 1**. See the Appendix for analogous diagrams for cases $n = 9, 10$, as well as tabulations of other empirical results.

3. Theoretical results

Although much of the following theory can be generalized from over fields to over arbitrary commutative rings, we will only work over fields.

3.1 Useful results for checking realizability

Proposition 2. *Let G be a graph on n vertices, let $\mathbb{F}$ be any field, and let $f(x) = x^n + a_{n-1}x^{n-1} + \cdots + a_0$ be a monic polynomial over $\mathbb{F}$. If $f(x) \in P(\mathbb{F}, G)$, then for all $\alpha \in \mathbb{F}$,*

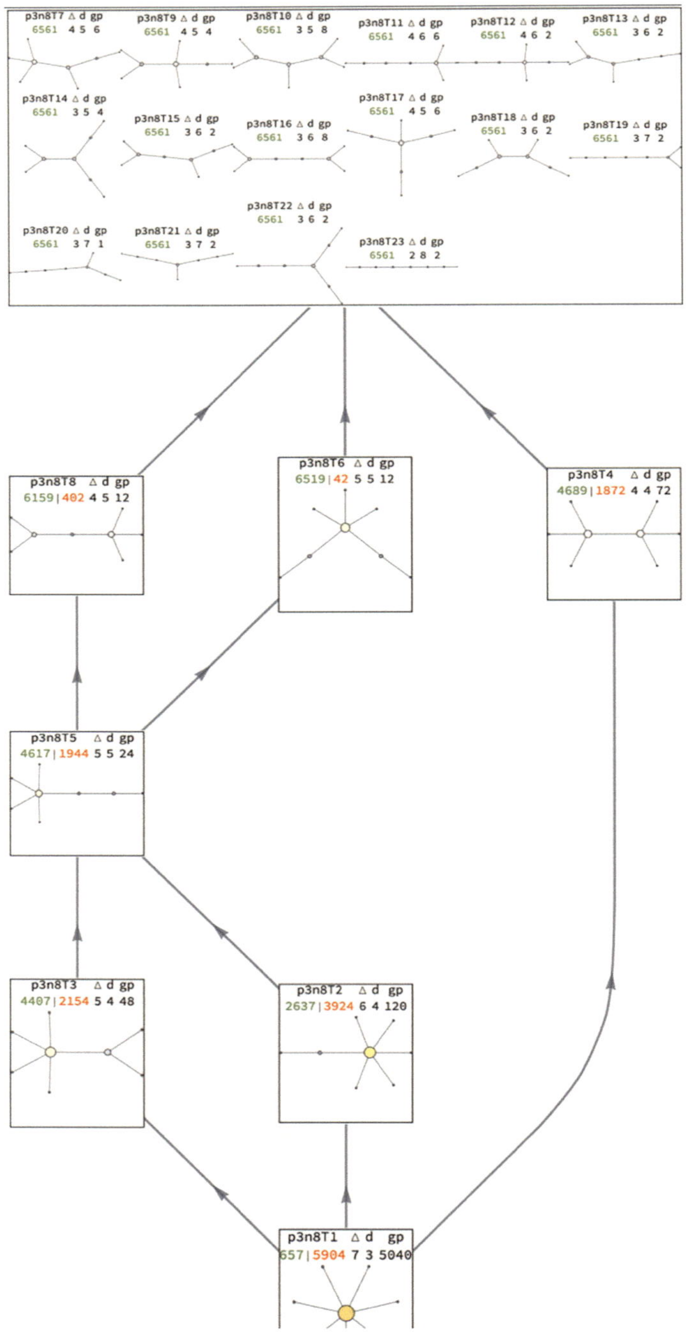

Figure 1.
The co-realizability partial ordering for all trees on $n = 8$ vertices over $\mathbb{F}_3$. The box containing a tree T contains information as follows. The green number is the number of polynomials T realizes, while the red is the number missed. $pn_1nn_2Tn_3$ indicates that the field is $\mathbb{F}_q = \mathbb{F}_{n_1}$, the tree is on n_2 vertices, and is the n_3th such tree to be indexed by Mathematica 13.1 function GraphData. *Δ, d, and* gp *indicates T's maximum degree, diameter, and the order of the graph automorpshim group of T, respectively.*

- $f(x-\alpha) \in P(\mathbb{F}, G)$, and
- $x^n + \alpha a_{n-1}x^{n-1} + \cdots + \alpha^n a_0 \in P(\mathbb{F}, G)$.

Proof: Let $f(x) \in P(\mathbb{F}, G)$. Then there exists $A \in \mathcal{F}(G)$ such that $p_A(x) = f(x)$. If $\alpha \in \mathbb{F}$, then $f(x+\alpha) = p_A(x-\alpha) = \det((x-\alpha)I - A) = \det(xI - (A+\alpha I))$. Since $A + \alpha I \in \mathcal{F}(G)$ whenever $A \in \mathcal{F}(G)$, the first point follows.

For the second point, recall that if $p_A(x) = x^n + \alpha a_{n-1}x^{n-1} + \cdots + \alpha^n a_0$, then $a_k = E_{n-k}(A)$ for all $k = 1, \dots, n$, where $E_k(A)$ denotes the sum of all k-by-k principal minors of A. The determinant of a k-by-k matrix is homogeneous of order k, so $E_k(\alpha A) = \alpha^k E_k(A)$.

3.2 Non-constructibility from concentrated pendants

The following theorem shows that given any graph G, we can construct a non-constructible graph over $\mathbb{F}_q$ by adding sufficiently many pendant vertices to a single fixed vertex of G.

Theorem 3. Let G be a graph on n vertices. If $k \geq q+1$ vertices are pendant at some vertex of G, then all members of $P(\mathbb{F}_q, G)$ have $k-q$ linear factors over $\mathbb{F}_q$, counting multiplicities.

Proof: Suppose $k \geq q+1$ vertices are pendant at a vertex v of G, and let $A \in \mathcal{F}(G)$. Then the characteristic matrix $A - xI$ is permutationally and then diagonally similar to a matrix of the form

$$A' - xI = \left[\begin{array}{cccccc|ccc}
a_1 - x & b_{1,2} & \cdots & b_{1,q} & \cdots & b_{1,k+1} & b_{1,k+2} & \cdots & b_{1,n} \\
1 & a_2 - x & & & & & & & \\
\vdots & & \ddots & & & & & & \\
1 & & & a_q - x & & & & & \\
\vdots & & & & \ddots & & & & \\
1 & & & & & a_{k+1} - x & & & \\
\hline
1 & & & & & & & & \\
\vdots & & & & & & & * & \\
1 & & & & & & & &
\end{array}\right].$$

There are only q elements of the field, so by relabeling if necessary, we can assume $a_q = a_{q+1}, a_{q+2} = a_{q+3}, \dots, a_k = a_{k+1}$. In addition, since $b_{1,k+1} \neq 0$, we can zero out the first row entries of columns $k+2, \dots, n$ by subtracting multiples of the $(k+1)$st column. Thus, $A' - xI$ is similar to a matrix of the form

$$A'' - xI = \left[\begin{array}{cccccc|c} a_1 - x & b_{1,2} \cdots & \cdots b_{1,q} \cdots & \cdots b_{1,k+1} & & & 0 \cdots \cdots 0 \\ 1 & a_2 - x & & & & & \\ \vdots & & \ddots & & & & \\ 1 & & & a_q - x & & & \\ \vdots & & & & \ddots & & \\ 1 & & & & & a_{k+1} - x & \\ \hline 1 & & & & & & \\ \vdots & & & & & & *' \\ 1 & & & & & & \end{array}\right].$$

$p_A(x) = p_{A''}(x)$ and $A'' - xI$ is a lower triangular block matrix, so

$$p_A(x) = p_{A''}(x) = \det(A'' - xI) = \det(S)\det[*'], \tag{3}$$

where S denotes the upper-left block of $A'' - xI$ shown above. The graph of S is a simple star on $k + 1$ vertices, so by Theorem 2.1 of [5], we know the characteristic polynomial of S has $\geq k - q$ roots over $\mathbb{F}_q$ since $k - q$ values are repeated on the diagonal. Then the result follows from the observation from Eq. (3) that $\det(S)$ divides $p_A(x)$.

3.3 Failure of the converse to Theorem 3

There is a counterexample to the converse of Theorem 3.

Proposition 4. *The converse of Theorem 3 is not true in general.*

Proof of Proposition 4: Let T be the tree in **Figure 2**, which has no vertices with more than $q = 3$ pendants. We claim each $f(x) \in P(\mathbb{F}_3, T)$ has a root over $\mathbb{F}_3$. Let $f(x) \in P(\mathbb{F}_3, T)$. Then there exists some $A \in N(T)$ such that $f(x) = \det(A - xI)$. Since any $A \in N(T)$ takes the form

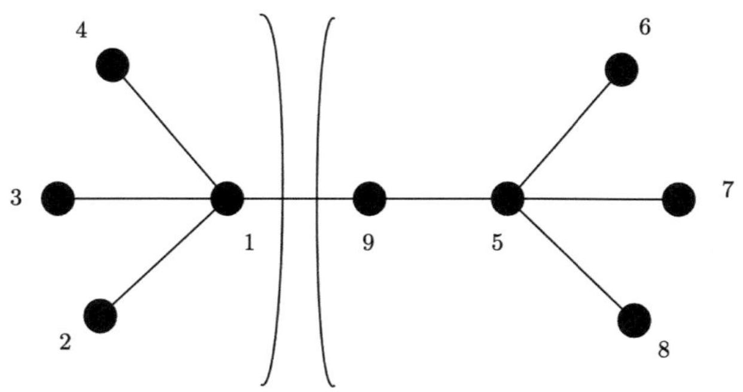

Figure 2.
The tree T on 9 vertices such that each $f(x) \in P(\mathbb{F}_3, T)$ has a root over $\mathbb{F}_3$, but no vertices of T have more than $q = 3$ pendants.

$$A = \begin{bmatrix} a_{1,1} & a_{1,2} & a_{1,3} & a_{1,4} & & & & & a_{1,9} \\ 1 & a_{2,2} & & & & & & & \\ 1 & & a_{3,3} & & & & & & \\ 1 & & & a_{4,4} & & & & & \\ & & & & a_{5,5} & a_{5,6} & a_{5,7} & a_{5,8} & a_{5,9} \\ & & & & 1 & a_{6,6} & & & \\ & & & & 1 & & a_{7,7} & & \\ & & & & 1 & & & a_{8,8} & \\ 1 & & & & 1 & & & & a_{9,9} \end{bmatrix},$$

we have $f(x)$ is the determinant of the matrix given by

$$\begin{bmatrix} a_{1,1}-x & a_{1,2} & a_{1,3} & a_{1,4} & & & & & a_{1,9} \\ 1 & a_{2,2}-x & & & & & & & \\ 1 & & a_{3,3}-x & & & & & & \\ 1 & & & a_{4,4}-x & & & & & \\ & & & & a_{5,5}-x & a_{5,6} & a_{5,7} & a_{5,8} & a_{5,9} \\ & & & & 1 & a_{6,6}-x & & & \\ & & & & 1 & & a_{7,7}-x & & \\ & & & & 1 & & & a_{8,8}-x & \\ 1 & & & & 1 & & & & a_{9,9}-x \end{bmatrix}.$$

It is left as a straightforward exercise to show the determinant of this matrix always has a linear factor.

3.4 Application of geometric Parter-Wiener, etc. theory

We initially suspected that if T is a graph on $n \leq |\mathbb{F}|$ vertices, then T is constructible. However, we found a counterexample by considering the simple star on four vertices over the field $\mathbb{F}_4$. To show that this indeed presents a counterexample, we apply methods from the relatively young Geometric Parter-Wiener, etc. theory [6]. The notation and terminology of this subsection is adopted from [6]. See also [3, Chapter 12].

Proposition 5. *There may be trees on q vertices that are non-constructible over* $\mathbb{F}_q$. *Indeed,*

$$\prod_{\alpha \in \mathbb{F}_4} (x - \alpha) = x^4 - x \notin P(\mathbb{F}_4, S_4). \tag{4}$$

Proof: Let v be the central vertex and u_1, u_2, u_3 the pendants at v. Let $k \in \{u_1, u_2, u_3\}$ and $\mathrm{gm}_A(k) = 1$. v is g-Parter for k—indeed, this follows from geometric Parter-Wiener, etc. theory since v is the only vertex with degree ≥ 2. In other words, if $k \in \sigma(A)$ and $k \in \{u_1, u_2, u_3\}$ then k appears at least least twice in $\{u_1, u_2, u_3\}$.

Suppose each $k \in \mathbb{F}_4$ is a simple eigenvalue of $A \in \mathcal{F}(S_4)$. Then $\mathrm{gm}_A(k) = 1$, so by the above statement, each of $\{u_1, u_2, u_3\}$ appears twice in $\{u_1, u_2, u_3\}$. The only way for this to happen is if $u_1 = u_2 = u_3 = k$, in which case $\mathrm{rank}(A - kI) \leq 2$. But then

$\mathrm{gm}_A(k) = \dim\ker(A - kI) \geq 2 > \mathrm{am}_A(k) = 1$, a contradiction. This establishes Eq. (4) and completes the proof.

Remark 6. In fact, $x^4 - x$ is the only polynomial over $\mathbb{F}_4$ that is not realized by S_4, which is to say $P(\mathbb{F}_4, S_4) = P_4(\mathbb{F}_4)\backslash\{x^4 - x\}$.

3.5 Algebraic branch duplication

This subsection largely adopts the definitions and background from [7, Section 2]. That work notes that many of these arguments also work for general graphs, but presents them in the case of trees, and we will take the same approach. To read more on the technique of branch duplication, one can also consult [3, Section 6.3].

Let T be a tree and $\{v, u_1\}$ be an edge of T. Let T_v (resp. T_1) be the connected component of T resulting from deletion of u_1 (resp. v) and containing v (resp. u_1). Denote by $n_i = |T_i|$ the number of vertices in T_i. An s-**combinatorial branch duplication** of T_1 at v is the tree obtained by appending $s \geq 1$ copies of the branch T$_1$ T at v. We will denote by $\mathbf{e}_{n_i,n_j}$, or sometimes $\mathbf{e}_{i,j}$ when clear from context, the n_i-by-n_j matrix over $\mathbb{F}$ with 1 as the $(1,1)$-entry and 0 as all remaining entries. We will also use the notation $\mathbf{e}_{vu}$ for vertices u and v of T, which is defined similarly. These notations may also be combined to yield $\mathbf{e}_{1,v}$ and $\mathbf{e}_{v,1}$, whose definitions should also be clear from context.

Let $A \in \mathcal{F}(T)$. By permutation similarity, A is similar to a matrix of the form

$$\left[\begin{array}{c|c} A[T_v] & a_{vu_1}\mathbf{e}_{v,1} \\ \hline a_{u_1v}\mathbf{e}_{1,v} & A[T_1] \end{array}\right] \tag{5}$$

where a_{vu_1}, a_{u_1v} are the nonzero entries of A correspond to the edge $\{v, u_1\}$ in T. Since the characteristic polynomial is invariant under matrix similarity, we may assume without loss of generality that A is of the form in (5) above.

Let T be a tree and v a vertex of T. Let $\breve{T}$ be the s-combinatorial branch duplication of the branch T_1 of at v. Let $u_2, \ldots, u_{1+s}$ (resp. $T_2, \ldots, T_{1+s}$) be the new neighbors of v (resp. the new branches at v) in $\breve{T}$.

We say that a matrix $\breve{A} = (\breve{a}_{ij})$ is s-**algebraic branch duplication** of A by $A[T_1]$ at v if $\breve{A}$ satisfies the following requirements:

i. $\breve{A}[T_v] = A[T_v]$ and $\breve{A}[T_1] = \cdots = \breve{A}[T_{1+s}] = A[T_1]$.

ii. $\breve{a}_{vu_i}\breve{a}_{u_iv} \neq 0$ $(i = 1, \ldots, 1+s)$, and $\breve{a}_{vu_1}\breve{a}_{u_1v} + \cdots + \breve{a}_{vu_{1+s}}\breve{a}_{u_{1+s}v} = a_{vu_1}a_{u_1v}$.

iii. The graph of $\breve{A}$ is $\check{T}$.

These conditions together imply

$$\breve{A} = \begin{bmatrix} A[T_v] & \breve{a}_{vu_1}\mathbf{e}_{v,1} & \breve{a}_{vu_2}\mathbf{e}_{v,1} & \cdots & \breve{a}_{vu_{1+s}}\mathbf{e}_{v,1} \\ \breve{a}_{u_1v}\mathbf{e}_{1,v} & A[T_1] & & & \\ \breve{a}_{u_2v}\mathbf{e}_{1,v} & & A[T_1] & & \\ \vdots & & & \ddots & \\ \breve{a}_{u_{1+s}v} & & & & A[T_1] \end{bmatrix}. \tag{6}$$

It turns out that $\breve{A}$ has a nice characteristic polynomial in relation to that of A:

Theorem 7. *([7], Theorem 1]). Let T be a tree, v a vertex of T, T_1 a branch of T at v and A be a combinatorially symmetric matrix whose graph is T. If* $\breve{A}$ is obtained from A by an s-algebraic branch duplication of summand $A[T_1]$ at v then $\breve{A}$ is similar to the block diagonal matrix $A \oplus_{i=1}^{s} A[T_1]$. Therefore

$$p_{\breve{A}}(t) = p_A(t) \cdot \left[p_{A[T_1]}(t)\right]^s.$$

and, for each eigenvalue λ of $\breve{A}$, we have

$$\mathrm{am}_{\breve{A}}(\lambda) = \mathrm{am}_A(\lambda) + s \cdot \mathrm{am}_{A[T_1]}(\lambda) \text{ and } \mathrm{gm}_{\breve{A}}(\lambda) = \mathrm{gm}_A(\lambda) + s \cdot \mathrm{gm}_{A[T_1]}(\lambda).$$

See [7] for a proof of Theorem 7, where it is noted that the proof works over all fields not equal to $\mathbb{F}_2$. Indeed, by the following lemma, proven in [8], we can find scalars satisfying condition (ii) above in every field other than $\mathbb{F}_2$:

Lemma 8. *([8], Lemma 2.3]). For any field $\mathbb{F} \neq \mathbb{F}_2$, any element $a \in \mathbb{F}$, and any integer $s \geq 2$, there exists $\breve{a}_1, \ldots, \breve{a}_s \in \mathbb{F}$ such that $\breve{a}_1 + \cdots + \breve{a}_s = a$.*

Corollary 9. *Let $\mathbb{F} \neq \mathbb{F}_2$ be a field and suppose T is a s-combinatorial branch duplication of a tree T_0 of the branch T_1 at a vertex v of T. Then for all $g(x) \in P(\mathbb{F}, T_1)$ and $h(x) \in P(\mathbb{F}, T_0)$, $[g(x)]^s h(x) \in P(\mathbb{F}, T)$.*

Remark 10. Using the same notation and hypotheses of Corollary 9, we can now construct a matrix $A \in \mathcal{F}(T)$ such that $p_A(x) = [g(x)]^s h(x)$ as follows: Let $B \in \mathcal{F}(T_1)$ and $C \in \mathcal{F}(T_0)$ be matrices such that $p_B(x) = g(x)$ and $p_C(x) = h(x)$. Define a new matrix C' by $C' = C + e_{v,u_1}\left(\breve{b}_0 - c_{v,u_1}\right)$. Then choose nonzero elements $\breve{b}_2, \ldots, \breve{b}_{1+s} \in \mathbb{F}$ such that $\breve{b}_1 + \breve{b}_2, \ldots, \breve{b}_{1+s} = c_{v,u_1}$. Then, where $\mathbf{e}_{1,0}$ is the $|T_1|$-by-$|T_0|$ matrix with a 1 in the (1, 1)-entry and zeros elsewhere, the matrix

$$A = \begin{bmatrix} C' & \breve{b}_1 \mathbf{e}_{0,1}^t & \cdots & \breve{b}_s \mathbf{e}_{0,1}^t \\ \mathbf{e}_{1,0} & B & & \\ \vdots & & \ddots & \\ \mathbf{e}_{1,0} & & & B \end{bmatrix}$$

satisfies $p_A(x) = [g(x)]^s h(x)$.

4. Conjectures

Let $\mathbb{F}$ be any field other than $\mathbb{F}_2$. A primary product of our empirical work is to reveal potentially valid statements of which we may not have otherwise thought. Here we record the most interesting of these statements, and a few relations among them, in hopes that they may spark an idea in others. Related conjectures are grouped together. In some cases, the evidence is very compelling. We give these as a list with comments.

1. If the simple star on n vertices realizes a monic polynomial p over $\mathbb{F}_q$, then all trees on n vertices realize p. This statement is likely true for infinite fields as well.

2. If every vertex in a tree T has degree strictly less than $\mathbb{F}$, then T is constructible.

   [Note: the same statement but with "strictly less than" replaced with "at most" is false, which can be seen by the counterexample used for Proposition 4, namely the 10-vertex nonlinear tree over $\mathbb{F}_3$. Also, since nonlinear trees are sources of frequent counterexamples (see, for example, [3]), it would not be surprising if the latter statement holds when restricted to linear trees.]

   a. An important special case is the path on n vertices. This has long been conjectured, though it has proven elusive. This was investigated extensively in [9], which partly inspired our work and is where this conjecture first arose.

3. When trees are categorized by diameter, those with larger diameters tend to realize more polynomials and are more often constructible than those with shorter diameters.

   a. Suppose a tree T is constructible over a field $\mathbb{F}$ and a new vertex is added to T, as to increase the diameter and produce T', then T' is also constructible.

   b. Suppose a tree T is non-constructible over a field $\mathbb{F}$ and a vertex is added to T to produce T' such that T' have the same diameter as T. Then T' is also non-constructible.

4. If a non-constructible tree T has at most $|\mathbb{F}|$ pendants at any vertexvertex, then sufficiently many vertices may be added to increase the diameter and achieve constructibility.

   [Note: This statement is motivated by the case $\mathbb{F} = \mathbb{F}_3$, as the data for that case suggests it. However, this statement may also hold in general.]

   a. Let $\mathbb{F}_q \neq \mathbb{F}_2$ be a finite field and let T be the tree produced by adding a pendant path to the simple star on $|\mathbb{F}_q| + 2$ vertices. Then $P(\mathbb{F}_q, T)$ is the set of monic polynomials over $\mathbb{F}_q$ with a root.

      [Note: Theorem 3 implies that $\mathrm{P}(\mathbb{F}_q, \mathrm{T})$ is a subset of the set of monic polynomials without a root, so to prove this it would be enough to prove the reverse inclusion.]

5. As the number of vertices increases, the fraction of polynomials that are constructible for any tree among all polynomials tends to 0.

   [Note: This is implied by the first conjecture, so an avenue to showing this would show that first.]

6. The number of co-realizability classes for trees on n vertices over $\mathbb{F}_3$ is 2^{n-4}.

7. If a graph G is constructible over $\mathbb{F}_q$, then the graph G' obtained by adding an edge to G is also constructible. In particular, for each positive integer n, there

exists a threshold number of edges, t_n such that any graph on n vertices with at least t_n edges is constructible.

8. K_n, the complete graph on n vertices, is constructible for all $n \geq 1$ over all fields other than $\mathbb{F}_2$.

   [We suggest that a proof for this statement be as follows: First, write the companion matrix for the desired polynomial. Then perform elementary operations to obtain similar matrices with properly more nonzero entries after each iteration. This may not be easy, but we believe it can be done; this holds for the real numbers, and the data supports it for small finite fields other than $\mathbb{F}_2$.]

5. Conclusion

Prior to this work, little was known. By studying the computational data, we formulated several conjectures and proved several. We demonstrated applications of branch duplication and geometric Parter-Wiener, etc. theory to the inverse characteristic polynomial problem by using them to prove results for general graphs over arbitrary finite fields. For some graphs, such as the path, the data suggests that all monic polynomials are realized over any field with more than two elements, although proving this has proved elusive. The authors hope the given results, arguments, and list of conjectures in this work will pave the way for future study.

The authors suggest that further work be done on the inverse characteristic polynomial problem for the case of the path over finite fields because a solution to this problem would likely result in a method for constructing a matrix with a given characteristic polynomial. Such construction techniques could also be pursued by considering methods such as the generalized partial fraction decomposition for arbitary fields (See [10]). This is motivated by the successful application of the partial fraction decomposition in [2] for matrices over the real numbers. The generalized Euclidean algorithm and Bézout's identity could also be investigated in the case of the path.

A. Appendix

We now show selected figures that were generated from this data. To download a zipped folder containing the raw data, some figures, and other information, please

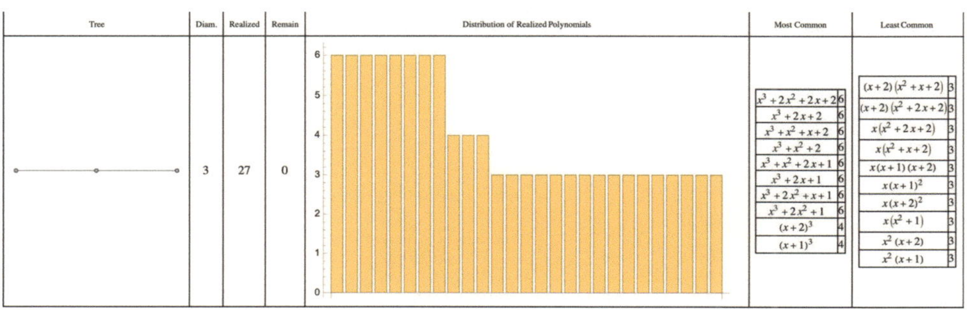

Figure 3.
Each bar in the bar chart corresponds to a polynomial in $P_3(\mathbb{F}_3)$. The bar's height at a given polynomial is the number of times that polynomial is realized by matrices (whose nonzero subdiagonal entries all equal 1) whose graph is the given tree.

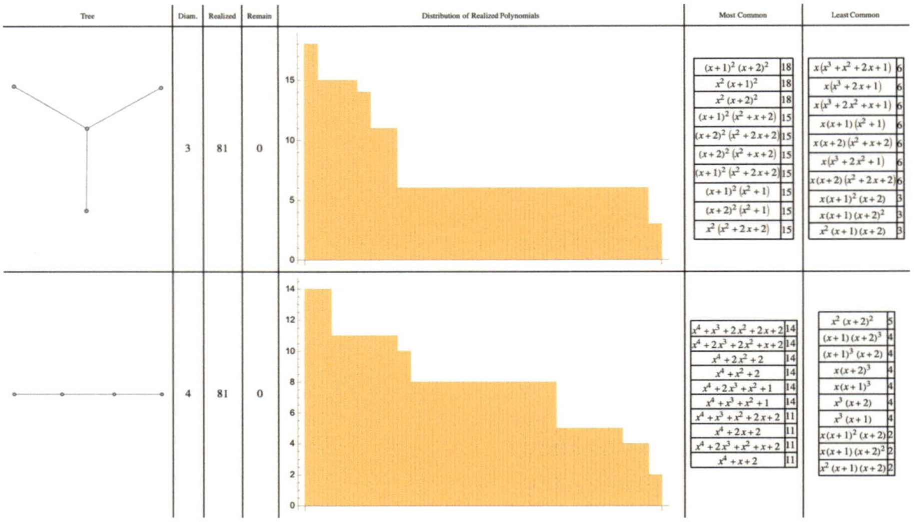

Figure 4.

Values on the x-axis of the bar chart represent different the polynomials of $P_4(\mathbb{F}_3)$. The bar height at a given polynomial indicates the number of times that polynomial is realized by matrices (whose nonzero subdiagonal entries all equal 1) whose graph is the given tree.

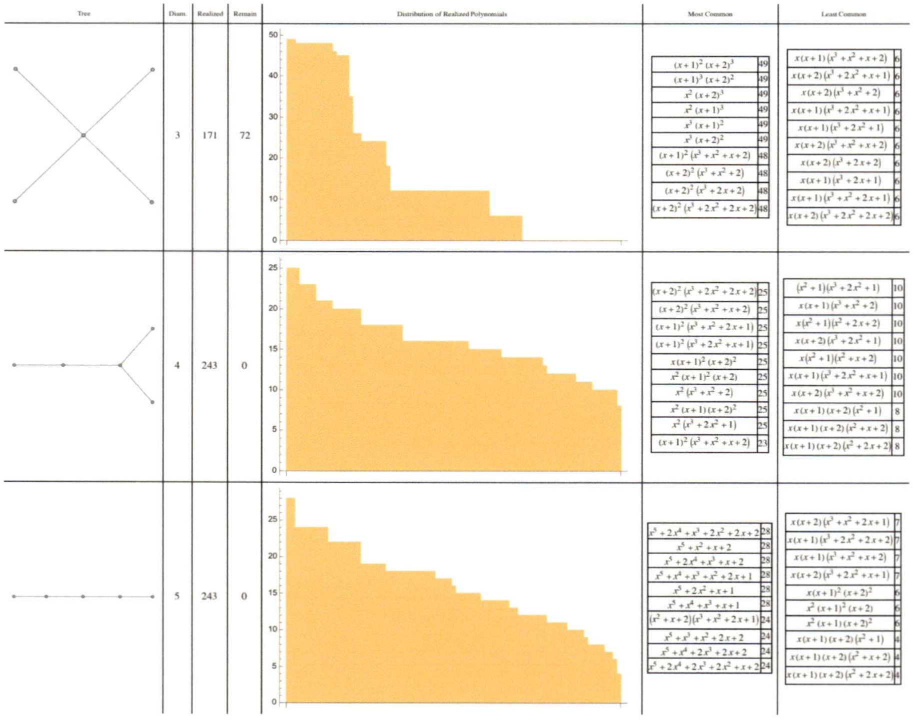

Figure 5.

Values on the x-axis of the bar chart represent different polynomials of $P_5(\mathbb{F}_3)$. The bar height at a given polynomial indicates the number of times that polynomial is realized by matrices (whose nonzero subdiagonal entries all equal 1) whose graph is the given tree.

DOI: http://dx.doi.org/10.5772/intechopen.1001164

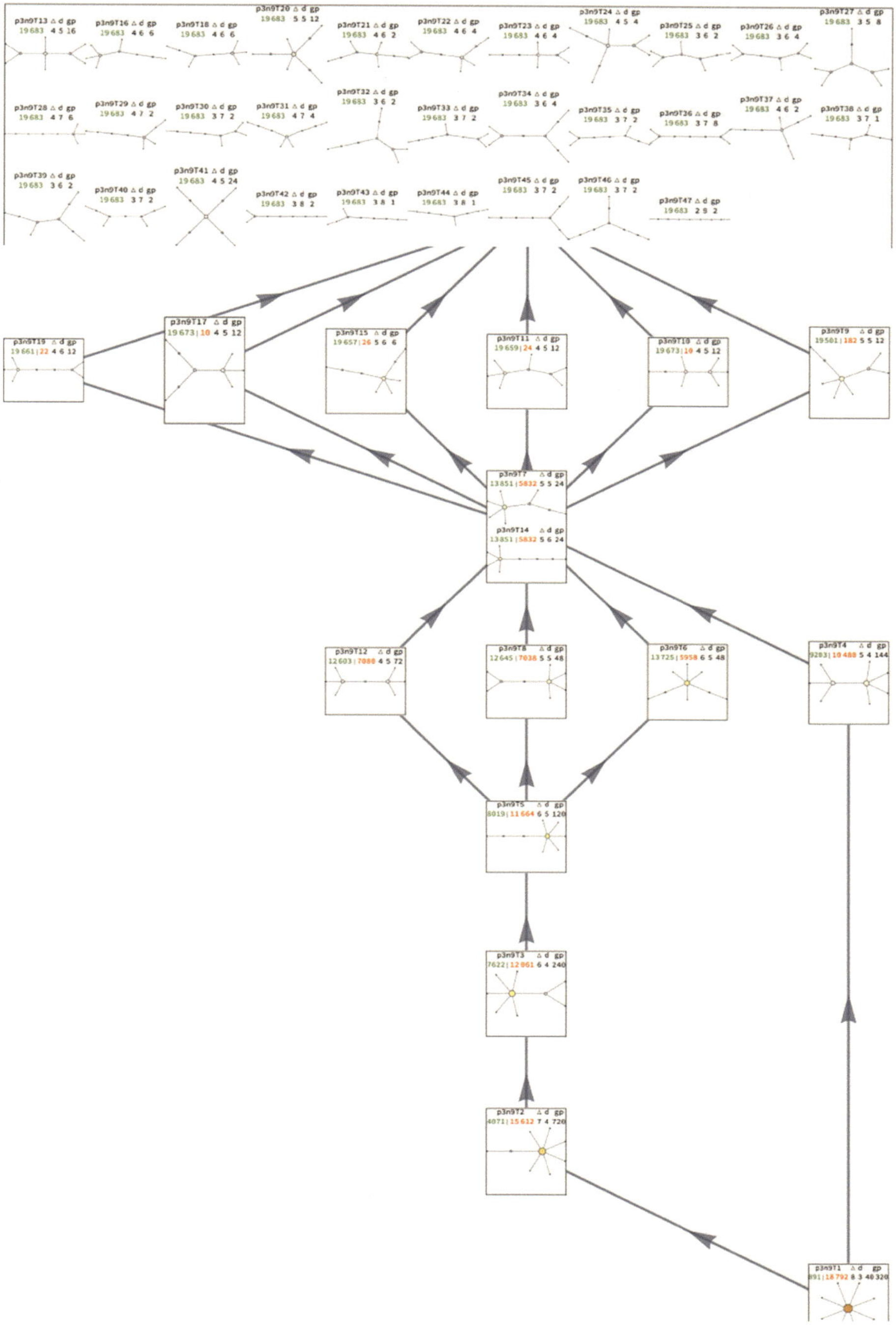

Figure 6.
The co-realizability partial ordering for all trees on $n = 9$ vertices over $\mathbb{F}_3$. The labeling scheme used is that of ***Figure 1***.

email the corresponding author or visit https:// drive.google.com/drive/u/1/folders/1m0t6fmBpGx6PMeKhw7UtCk6Hg8gf2e1k. The folder also contains diagrams analogous to **Figures 3** and **4** through $n = 10$ vertices, together with higher-quality versions of these figures (**Figures 5** and **6**).

Author details

Charles R. Johnson[1], Greyson C. Wesley[2*], Xiaoyu Lin[1], Xiwen Liu[1] and Sihan Zhou[1]

1 College of William and Mary, Williamsburg, USA

2 University of Notre Dame, Notre Dame, USA

*Address all correspondence to: gwesley@nd.edu

References

[1] Friedland S. Inverse eigenvalue problems. Linear Algebra and its Applications. 1977;**17**:15-51

[2] Gruner E, Johnson CR. The inverse characteristic polynomial problem for trees. In: Operator Theory, Functional Analysis and Applications. Cham, Switzerland: Birkhäuser; 2021. pp. 329-351

[3] Johnson CR, Saiago CM. Eigenvalues, Multiplicities and Graphs. Cambridge Tracts in Mathematics. Cambridge, New York: Cambridge University Press; 2018

[4] Wolfram Research Inc. Mathematica, Version 13.1. Champaign, IL; 2023

[5] Johnson CR, Leal-Duarte A. Complete spectral theory for matrices over a field whose graph is a star. Manuscript. 2022

[6] Johnson CR, Barsotti J, Shelburne E. Characteristic polynomials of irreducible tridiagonal matrices over finite fields. Manuscript. 2020

[7] Saiago CM. Diagonalizable matrices whose graph is a tree: The minimum number of distinct eigenvalues and the feasibility of eigenvalue assignments. Special Matrices. 2019;**7**:316-326

[8] Johnson CR, Saiago CM. Geometric Parter–Wiener, etc. theory. Linear Algebra and its Applications. 2018;**537**: 332-347

[9] Adkins W, Davidson M. Synthetic partial fraction decompositions. Mathematics Magazine. 2008;**81**:16-26

[10] Horn RA, Johnson CR. Matrix Analysis. 2nd ed. Cambridge, New York: Cambridge University Press; 2013

Chapter 7

On Computing of Independence Polynomials of Trees

Ohr Kadrawi, Vadim E. Levit, Ron Yosef and Matan Mizrachi

Abstract

An independent set in a graph is a set of pairwise nonadjacent vertices. Let $\alpha(G)$ denote the cardinality of a maximum independent set in the graph $G = (V, E)$. In 1983, Gutman and Harary defined the independence polynomial of G $I(G; x) = \sum_{k=0}^{\alpha(G)} s_k x^k = s_0 + s_1 x + s_2 x^2 + \ldots + s_{\alpha(G)} x^{\alpha(G)}$, where s_k denotes the number of independent sets of cardinality k in the graph G. A comprehensive survey on the subject is due to Levit and Mandrescu, where some recursive formulas are allowing computation of the independence polynomial. A direct implementation of these recursions does not bring about an efficient algorithm. In 2021, Yosef, Mizrachi, and Kadrawi developed an efficient way for computing the independence polynomials of trees with n vertices, such that a database containing all of the independence polynomials of all the trees with up to $n - 1$ vertices is required. This approach is not suitable for big trees, as an extensive database is needed. On the other hand, using dynamic programming, it is possible to develop an efficient algorithm that prevents repeated calculations. In summary, our dynamic programming algorithm runs over a tree in linear time and does not depend on a database.

Keywords: independent set, independence polynomial, tree decomposition, dynamic programming, post-order traversal

1. Introduction

1.1 Definitions

This paper $G = (V, E)$ is a simple (i.e., a finite, undirected, loop less, and without multiple edges) graph with vertex set $V = V(G)$ of cardinality $|V(G)| = n(G)$ and edge set $E = E(G)$ of cardinality $|E(G)| = m(G)$. The *neighborhood* of a vertex $v \in V$ is the set $N_G(v) = \{u : u \in V \text{ and } uv \in E\}$, and $N_G[v] = N_G(v) \cup v$.

The *disjoint union* of the graphs G_1 and G_2 is the graph $G = G_1 \cup G_2$ having as vertex set the disjoint union of $V(G_1), V(G_2)$, and as edge set the disjoint union of $E(G_1), E(G_2)$.

The *tree decomposition* of a graph G is a tree T of "bags," where if edge $(u, v) \in E(G)$ then u and v are together in same "bag," and $\forall w \in V(G)$ the "bags" containing w are connected in T. The *width* of a tree decomposition is equal to one less than the maximum bag size and the *treewidth* of G equal to the least width of all tree decompositions for G.An *independent set* or a *stable set* in G is a set of pairwise nonadjacent vertices. An independent set of maximum size will be referred to as a *maximum independent set* of G, and the *independence number* of G, denoted by $\alpha(G)$, is the cardinality of a maximum independent set in G.

Let s_k be the number of independent sets of cardinality k in a graph G. For example, $s_0 = 1$ is the number of independent sets of minimum cardinality in G (i.e, the number of empty sets). The polynomial.

$I(G;x) = \sum_{k=0}^{\alpha(G)} s_k x^k = s_0 + s_1 x + s_2 x^2 + \ldots + s_{\alpha(G)} x^{\alpha(G)}$, is the *independence polynomial* of G [1], the *independent set polynomial* of G [2], or the *stable set polynomial* of G [3]. Some updated observations concerning the independence polynomial may be found in [4, 5].

A finite sequence of real numbers $(a_0, a_1, a_2, \ldots, a_n)$ is said to be:

- *unimodal* if there is some $k \in 0, 1, \ldots, n$, called the *mode* of the sequence, such that

$$a_0 \leq \ldots \leq a_{k-1} \leq a_k \geq a_{k+1} \geq \ldots \geq a_n; \tag{1}$$

  the mode is *unique* if $a_{k-1} < a_k > a_{k+1}$;

- *logarithmically concave* (or simply, *log-concave*) if the inequality

$$a_k^2 \geq a_{k-1} \cdot a_{k+1} \tag{2}$$

  is valid for every $i \in 1, 2, \ldots, n-1$.

It is known that any log-concave sequence of positive numbers is also unimodal. Unimodal and log-concave sequences occur in many areas of mathematics, including algebra, combinatorics, graph theory, and geometry.

For instance:

- The sequence of binomial coefficients, presented in the nth row of Pascal's triangle is log-concave.
- Let us consider a sequence of vector spaces $V_0, V_1, \ldots V_n$ and the corresponding linear transformations $\phi_k : V_k \to V_{k+1}, 0 \leq k \leq \lfloor (n-1)/2 \rfloor$. If dim $V_i = a_i$ for $0 \leq i \leq n$, the mappings ϕ_k are injective for $0 \leq k \leq \lfloor (n-1)/2 \rfloor$, and $V_i = V_{n-i}$, then the sequence $a_0, a_1, \ldots, a_n$ is palindromic and unimodal [6].
- Let (P, ω) be a labeled poset. For $s \in \mathbb{N}$, let $\Omega((P, \omega); s)$ be the number of (P, ω)-partitions with the largest part $\leq s$. Then the sequence $\Omega((P, \omega); s)_{s \in \mathbb{N}}$ is log-concave [7, 8].
- Let (W, S) be a finite Coxeter system with $d(x) = |s \in S : l(xs) < l(x)|$. Then the polynomial, $\sum_{x \in W} q^{d(x)}$ is palindromic and unimodal [7].

In addition, see Refs. [9–12], and especially, the surveys of Brenti [7] and Stanley [6].

A considerable amount of literature has been published on the unimodality and the log-concavity of various polynomials defined on graphs. In the context of our paper, for instance, it is worth mentioning the following results:

- A spider is a tree with one vertex of degree at least 3 and all others with degree at most 2. The well-covered spider S_n, $n \geq 2$ has one vertex of degree $n + 1$, n vertices of degree 2, and $n + 1$ vertices of degree 1. The independence polynomial of any well-covered spider is unimodal [13].
- If a_k denotes the number of matchings of size k in a graph, then the sequence of these numbers is unimodal [14].
- The independence polynomial of a claw-free graph is log-concave, and, hence, unimodal, as well [15].
- A famous result by Chudnovsky and Seymour stated that all the roots of $I(G;x)$ are real whenever G is a claw-free graph, which also proves the log-concavity and, consequently, the unimodality of $I(G;x)$ for all claw-free graphs G [16].
- The domination polynomial of almost every graph is unimodal [17].
- For any graph, the numbers of dependent k-sets form a log-concave sequence (dependent set is a set that is not independent) [18].

Alavi, Malde, Schwenk, and Erdös conjectured that independence polynomials of trees are unimodal [19]. Yosef, Mizrachi, and Kadrawi developed an approach for computing the independence polynomials of trees on n vertices using a database [20]. This approach is based on an extensive database containing the independence polynomials of all the trees with up to $n - 1$ vertices, and the induction step computing the independence polynomials of all the trees with n vertices based on their $n - 1$ counterparts. To this end, it supports the unimodality and log-concavity of independence polynomials of trees with up to 20 vertices. Further, Radcliffe has verified that independence polynomials of trees up to 25 vertices are log-concave [21].

In 2004, Levit and Mandrescu conjectured that every forest is log-concave [22]. In 2011, Galvin added a comment that for a tree/forest/bipartite graph, the unimodality conjecture may be strengthened to the corresponding log-concavity one [23].

1.2 Computing the independence polynomial

To compute independence polynomials $I(G;x)$, as shown in the survey [24] and also in ref. [1, 2, 25], one can use the following recursive formula:

$$I(G;x) = I(G - v;x) + x \cdot I(G - N[v];x). \tag{3}$$

To compute the independence polynomial of the union of disjoint graphs the formula reads as follows [2, 24]:

$$I(G_1 \cup G_2;x) = I(G_1;x) \cdot I(G_2;x). \tag{4}$$

1.3 Problem

Let us recall some known facts on similar problems:

- Computing $\alpha(G)$, the cardinality of a **maximum independent set** in a graph, is an NP-hard problem.
- Computing the **independence polynomial** of a graph is also an NP-hard problem since the degree of the independence polynomial is equal to $\alpha(G)$.
- There are families of graphs with computed closed-form expressions of their independence polynomials [26], but from what is known, general trees are not one of them.
- For some families of graphs, there are polynomial algorithms able to compute their $\alpha(G)$. Does it give us hope for a polynomial algorithm computing their independence polynomials?

According to Bodlaender [27], many practical problems rely heavily on graphs with bounded treewidth, for instance, trees/forests having treewidth ≤ 1. Tittmann [28] offers a way to construct an algorithm that computes the independence polynomial of a graph with bounded treewidth in polynomial time based on a specific partition. Starting from these considerations, we accurately developed the corresponding algorithm working on trees.

The algorithm presented in ref. [20] computes efficiently independence polynomials of small-sized trees, but really big trees may require an enormous database to support computing their independence polynomials.

2. Main idea of the algorithm

The new algorithm that we suggest for computing independence polynomials of trees does not require any database for its implementation. Instead, it used the dynamic programming technique.

The independence polynomial $I(T;x)$ represented in the algorithm as a list with $\alpha(G)+1$ cells in the following format:

$$\left[s_{k=\alpha(G)}, s_{k=\alpha(G)-1}, \dots, s_{k=1}, s_{k=0}\right]. \tag{5}$$

Examples of some graphs are shown in **Table 1**.

Graph	I(G; x)	Stored as
P_1	$x+1$	[1, 1]
P_2	$2x+1$	[2, 1]
P_3	x^2+3x+1	[1,3,1]

Table 1.
Path graphs with 1,2, and 3 vertices and their representation in the algorithm.

From the computing independence polynomial formula, as described in Eq. (3), we can see that every vertex can be calculated just after its children and grandchildren are calculated. Post-order-traversal validates that the children and grandchildren vertices will be computed before the father vertex.

In order to construct the algorithm, we set two base cases, and then the recursive process:

1. $|V| = 1$; **isolated vertex**:

   - $I(T - v; x) = [1]$ because in $T - v$ there are no vertices, so there is one independent set of cardinality zero (i.e., empty set).

   - $I(T - N[v]; x) = [1]$ from the same reason above. so:

$$I(T; x) = [1] + [1, 0] \cdot [1] = [1, 1]. \tag{6}$$

2. $|V| = 2$; **tree with two vertices,** P_2:

   - $I(T - v; x) = [1, 1]$ because when vertex v is removed, the graph stays with only one vertex, and this subgraph has been handled in the second bullet of the previous case.

   - $I(T - N[v]; x) = [1]$ because in $T - N[v]$ there are no vertices, so there is one independent set of cardinality zero (i.e., empty set). Thus:

$$I(T; x) = [1, 1] + [1, 0] \cdot [1] = [2, 1]. \tag{7}$$

3. $|V| > 2$:

   a. Start with traveling on the tree in the post-order traversal. When reaching a leaf node, like case 1, it can calculate by:

      - $I(T - v; x) = [1]$,

      - $x \cdot I(T - N[v]; x) = [1, 0]$,

      - $I(T; x) = [1, 1]$.

   b. For an inner vertex that all its children were calculated, when vertex v is removed, the number of connected components can rise, and in such case, computation of subgraphs union, as described in Eq. (4), is needed.

   So in purpose to calculate $I(T; x)$, compute next parameters:

$$I(T - v; x) = \Pi_{u \in children[v]} I(T; x), \tag{8}$$

      - $I(T - N[v]; x)$ parameter said that we remove the vertex v with its neighbors so we can use $I(T - v; x)$ parameter of the children:

- $I(T - N[v]; x) = \Pi_{u \in children[v]} I(T - u; x)$.

Finally, use Eq. (3) to calculate $I(T; x)$.

3. Main algorithm

The algorithm in IP (compute **i**ndependence **p**olynomial) function starts with two base cases: the minimal trees P_1 with only one vertex and P_2 with 2 vertices and an edge between them, and their independence polynomials are $x + 1$ and $2x + 1$, respectively as shown in **Table 1**. After that, select an inner vertex from the tree (that its degree is ≥ 2—more explanation in the next section) and set it to be the root node. Then we use IP_VISIT function in order to walk over the tree.

Algorithm 1: IP(T)

Input: Tree T as adjacency list
Output: A list I that represents the independence polynomial of T
// Two base cases:
  if $|V| == 1$ then
    ⌊ return *[1,1]*
  if $|V| == 2$ then
    ⌊ return [2,1]
  // Select a root vertex that is not a leaf:
  $root \leftarrow findInnerVertex()$
  // Call the recursive function:
  IP_VISIT(T, root)
  // Return the independence polynomial of T:
  return $I[root]$

IP_VISIT (like DFS_VISIT) function starts from some vertex and explores all the sub-tree from it. It goes as far as it can down for some branch until it reaches a leaf and then backtracks until it finds a new branch, and then explores it. The algorithm does this until the entire sub-tree has been explored.

In the algorithm, we use four lists:

- I - That stores I(T; x) for each calculated vertex
- V - That stores I(T-v; x) for each calculated vertex
- N - That stores I(T-N[v]; x) for each calculated vertex
- C - That stores the children's vertices for each calculated vertex

Algorithm 2: IP_VISIT(T, root)

```
   Input: Tree T as adjacency list, root vertex
   Output: Three lists: I = I(T; x), V = I(T-v; x), N = I(T-N[v]; x)
   // Stop condition:
1  if root is a leaf then
2  |  V[root] ← [1]
3  |  N[root] ← [1]
4  |  I[root] ← [1, 1]
5  else
   |  // Explore undiscovered vertices that are neighbors
   |     of root:
6  |  foreach vertex u that neighbor of v and never explored do
   |  |  // Save the hierarchy:
7  |  |  C[root].append(u)
   |  |  // Call again recursively with u:
8  |  |  IP_VISIT(T, u)
9  |  left ← [1] // Identity element to multiply
10 |  right ← [1,0] // Initialize as x
   |  // Calculate the union of sub-graphs
11 |  foreach u in C do
12 |  |  left ← left * I[u] // polynomial multiplication
13 |  |  right ← right * V[u]
   |  // Set all lists in root index
14 |  V[root] ← left
15 |  N[root] ← right
16 |  I[root] ← left + right // polynomial Addition
```

4. Proof of correctness

Lemma 4.1. *IP_VISIT(T, root) is called exactly once for each vertex in the graph.*

Proof: Clearly, IP_VISIT(T, root) is called for a vertex u only if it is not discovered. The moment it is called, IP_VISIT(T, root) cannot be called for vertex u again. Furthermore, because T is a connected component, and IP_VISIT(T, root) uses post-order traversal in the implementation, it ensures that it will be called for every vertex in T.

Lemma 4.2. *In the body of "for", each loop that explores undiscovered vertices that are root-neighbors (lines 6–8) is executed exactly once for each edge (v,u) in the graph.*

Proof: IP_VISIT(T, root) is called exactly once for each vertex root (Claim 1). And the body of the loop is executed once for all the unseen edges that connect to the root.

Corollary 4.3. *The algorithm can start from every vertex root such that.*

$deg(v) \geq 2$ and get the same runtime.

Proof: The order of walking on the tree is in post-order travels. In this way, it goes through each edge exactly twice so that no matter which vertex v ($deg(v) \geq 2$) it starts from, the running time will remain the same.

Therefore, the complexity of the IP_VISIT(T, root) algorithm is $O(n + m)$. Taking into account that our input is a tree, the complexity summarizes to $O(n)$.

5. Applications of the algorithm

As a part of the computations, in order to support the Alavi, Malde, Schwenk, and Erdös conjecture [19], using the linear algorithm described above, it was verified that for all trees up to 25 vertices, their independence polynomials are log-concave (and, consequently, unimodal), see also ref. [21]. All of the sudden, when the number of vertices of a tree reached 26, there were found two trees having their independence polynomials unimodal but not log-concave. In other words, these trees are counter-examples to the conjecture due to Levit, Mandrescu [22], and Galvin [23].

The independence polynomials of the trees T_1 and T_2 defined in **Figure 1** are as follows:

$$\begin{aligned} I(T_1;x) = x^{14} + 51x^{13} + 2979x^{12} + 18683x^{11} + 55499x^{10} + 100144x^9 + \\ 121376x^8 + 103736x^7 + 63933x^6 + 28551x^5 + 9142x^4 + 2040x^3 + \\ 300x^2 + 26x + 1, \end{aligned} \tag{9}$$

where the non-log-concavity is demonstrated by the coefficient of x^{13}: $51^2 < 2979$, and

$$\begin{aligned} I(T_2;x) = x^{14} + 48x^{13} + 2372x^{12} + 15498x^{11} + 48086x^{10} + 90178x^9 + \\ 112870x^8 + 98968x^7 + 62183x^6 + 28147x^5 + 9089x^4 + 2037x^3 + \\ 300x^2 + 26x + 1, \end{aligned} \tag{10}$$

where the non-log-concavity is demonstrated by the coefficient of x^{13}: $48^2 < 2372$.

6. Extensions of T1

T_1 from **Figure 1** is just an example belonging to an infinite family of trees such that their independent polynomials are not log-concave. Their structure, which we denoted *3,k,k structure*, is described in the following and drawn in **Figure 2**:

- the tree has one center, denoted v_0 that is connected to three vertices v_1, v_2, v_3;
- v_1 is connected to $K_2 \cup K_2 \cup K_2$;
- v_2 is connected to $K_2 \cup \ldots \cup K_2$, k times;
- v_3 is also connected to another $K_2 \cup \ldots \cup K_2$, k times.

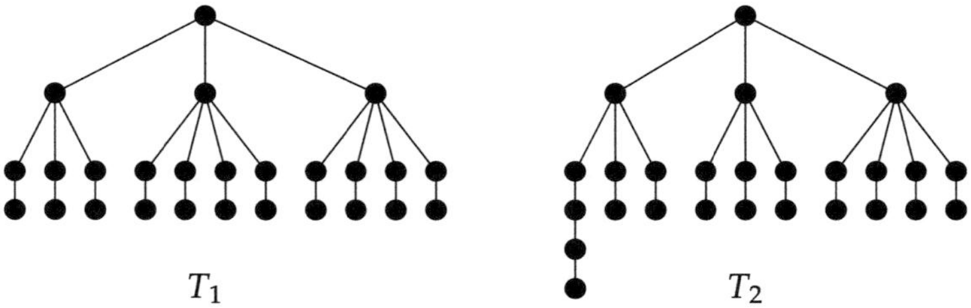

Figure 1.
Two trees with 26 vertices whose independence polynomials are not log-concave.

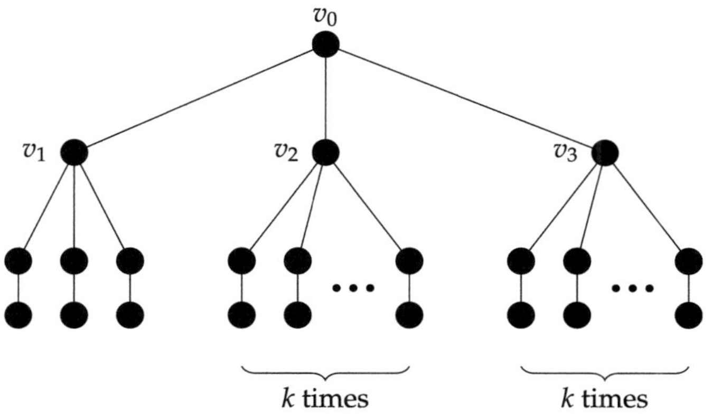

Figure 2.
An illustration of the 3,k,k structure of trees that have non-log-concave independence polynomials.

Lemma 6.1 *All trees of the* $3, k, k$ *structure, where* $k \geq 4$, *have non-log-concave independence polynomials.*

Proof: Let us compute the independence polynomial of a tree having $3, k, k$-structure, and choose v_0 to be the first vertex to remove from the graph.

$$I(G)_{v_0} = I(G - v_0) + x \cdot I(G - N[v_0]). \tag{11}$$

Expand the first term in that sum:

$$\begin{aligned} I(G - v_0) &= \left[(2x+1)^k + x(x+1)^k\right] \cdot \left[(2x+1)^k + x(x+1)^k\right] \\ &\quad \cdot \left[(2x+1)^3 + x(x+1)^3\right] \\ &= \left[\sum_{i=0}^{k} \binom{k}{i} (2x)^i + x \sum_{i=0}^{k} \binom{k}{i} x^i\right]^2 \cdot \left[(2x+1)^3 + x(x+1)^3\right] \\ &= \left[x^{k+1} + (2^k + k)x^k + \left(k \cdot 2^{k-1} + \frac{k(k-1)}{2}\right)x^{k-1} + \cdots\right]^2 \\ &\quad \cdot \left[x^4 + 11x^3 + 15x^2 + \cdots\right]. \end{aligned} \tag{12}$$

We can divide the first term expression into three factors $A \cdot B \cdot C$ such that:

$$\begin{aligned} A &= \left[x^{k+1} + (2^k + k)x^k + \left(k \cdot 2^{k-1} + \frac{k(k-1)}{2}\right)x^{k-1} + \cdots\right]; \\ B &= \left[x^{k+1} + (2^k + k)x^k + \left(k \cdot 2^{k-1} + \frac{k(k-1)}{2}\right)x^{k-1} + \cdots\right]; \\ C &= \left[x^4 + 11x^3 + 15x^2 + \cdots\right]. \end{aligned} \tag{13}$$

Expand the second term in that sum:

$$\begin{aligned}
I(G - N[v_0]) &= x \cdot (2x + 1)^{2k+3} \\
&= x \cdot \left[\sum_{i=0}^{2k+3} \binom{2k+3}{i} (2x)^i \right] \\
&= x \cdot \left[(2k)^{2k+3} + \cdots \right] \\
&= 2^{2k+3} x^{2k+4} + \cdots
\end{aligned} \tag{14}$$

The highest exponent that one can reach is $2k + 6$. We obtain it by taking the highest exponent from every factor, that is, from both factors A and B, we choose x^{k+1}, while from C factor we choose x^4, so

$$x^{n+1} \cdot x^{k+1} \cdot x^4 = X^{2k+6} \tag{15}$$

with the coefficient equal to 1.

To calculate the coefficient of x^{2k+5} we have three following options:

- multiply $(2^k + k)x^k$ from A, x^{k+1} from B, and x^4 from C,
- multiply x^{k+1} from A, $(2^k + k)x^k$ from B, and x^4 from C,
- multiply x^{k+1} from A, x^{k+1} from B, and $11x^3$ from C.

In such a way we obtain the following:

$$(2^k + k)x^k \cdot x^{k+1} \cdot x^4 + x^{k+1} \cdot (2^k + k)x^k \cdot x^4 + x^{k+1} \cdot x^{k+1} \cdot 11x^3 = (2^{k+1} + 2k + 11)x^{2k+5} \tag{16}$$

with the coefficient equal to $2^{k+1} + 2k + 11$.

To calculate the coefficient of x^{2k+4} we have six following options:

- multiply x^{k+1} from A, x^{k+1} from B, and $15x^2$ from C,
- multiply x^{k+1} from A, $(2^k + k)x^k$ from B, and $11x^3$ from C,
- multiply x^{k+1} from A, $\left(k \cdot 2^{k-1} + \frac{k(k-1)}{2}\right)x^{k-1}$ from B, and x^4 from C,
- multiply $(2^k + k)x^k$ from A, x^{k+1} from B, and $11x^3$ from C,
- multiply $(2^k + k)x^k$ from A, $(2^k + k)x^k$ from B, and x^4 from C,
- multiply $\left(k \cdot 2^{k-1} + \frac{k(k-1)}{2}\right)x^{k-1}$ from A, x^{k+1} from B, and x^4 from C.

Finally, we add the coefficient 2^{2k+3} from the second term. In such a way we obtain the following:

$$x^{k+1} \cdot x^{k+1} \cdot 15x^2 + x^{k+1} \cdot (2^k + k)x^k \cdot 11x^3 + x^{k+1} \cdot \left(k \cdot 2^{k-1} + \frac{k(k-1)}{2}\right) \cdot x^4$$
$$+ (2^k + k)x^k \cdot x^{k+1} \cdot 11x^3 + (2^k + k)x^k \cdot (2^k + k)x^k \cdot x^4 + \left(k \cdot 2^{k-1} + \frac{k(k-1)}{2} x^{k-1}\right)$$
$$\cdot x^{k+1} \cdot x^4 + 2^{2k+3}x^{2k+4}$$
$$= \left(2^{2k+3} + 2^{2k} + (22 + 3k)2^k + 2k^2 + 21k + 15\right)x^{2k+4}. \tag{17}$$

with the coefficient equal to $2^{2k+3} + 2^{2k} + (22 + 3k)2^k + 2k^2 + 21k + 15$.
Thus the independence polynomial is as follows:

$$I(G) = x^{2k+6} + (2^{k+1} + 2k + 11)x^{2k+5} + \left[2^{2k+3} + 2^{2k} + (22 + 3k)2^k + 2k^2 + 21k + 15\right]x^{2k+4} + \cdots \tag{18}$$

Now, let us prove the non-log-concavity in the x^{2k+5} term of the independence polynomial:

$$\left(2^{k+1} + 2k + 11\right)^2 < 1 \cdot \left[2^{2k+3} + 2^{2k} + (22 + 3k)2^k + 2k^2 + 21k + 15\right] \tag{19}$$

The left-hand side and the right-hand side are equal to $k \approx 3.5357$ so the left-hand side is smaller than the right-hand side from $k = 4$ and above.

7. Extensions of T2

T_2 from **Figure 1** is just an example belonging to an infinite family of trees whose independence polynomials are not log-concave. Let us denote the left sub-tree "3*." Their structure, which we denote *3*,k,k + 1 structure* is described in the following, and drawn in **Figure 3**:

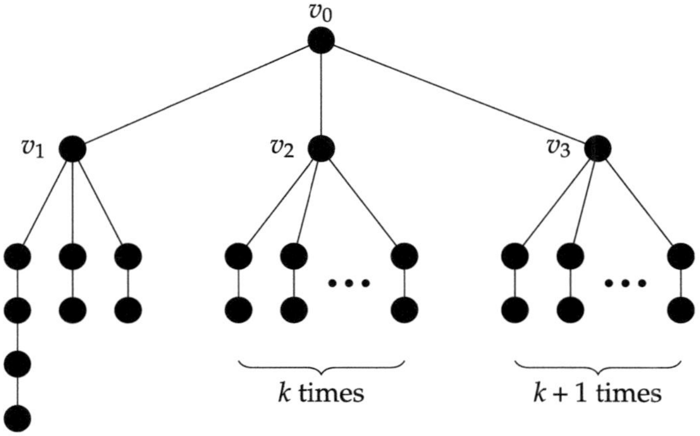

Figure 3.
An illustration of the 3,k,k + 1 structure of trees that have non-log-concave independence polynomial.*

the tree has one center, denoted v_0 that is connected to three vertices v_1, v_2, v_3;

- v_1 is connected to $P_4 \cup K_2 \cup K_2$;
- v_2 is connected to $K_2 \cup \ldots \cup K_2$, k times;
- v_3 is connected to another $K_2 \cup \ldots \cup K_2$, $k+1$ times.

Lemma 7.1. *All trees of the* $3^*, k, k+1$ *structure, where* $k \geq 3$, *have non-log-concave independence polynomials.*

Proof: Let us compute the independence polynomial of a tree having $3^*, k, k+1$ structure, and choose v_0 to be the first vertex to remove from the graph.

$$I(G)_{v_0} = I(G - v_0) + x \cdot I(G - N[v_0]) \tag{20}$$

Expand the first term in that sum:

$$\begin{aligned}
I(G - v_0) &= \left[(2x+1)^{k+1} + x(x+1)^{k+1}\right] \cdot \left[(2x+1)^{k} + x(x+1)^{k}\right] \\
&\quad \cdot \left[(2x+1)^2(3x^2+4x+1) + x(x+1)^2(x^2+3x+1)\right] \\
&= \left[\sum_{i=0}^{k+1}\binom{k+1}{i}(2x)^i + x\sum_{i=0}^{k+1}\binom{k+1}{i}x^i\right] \cdot \left[\sum_{i=0}^{k}\binom{k}{i}(2x)^i + x\sum_{i=0}^{k}\binom{k}{i}x^i\right] \\
&\quad \cdot \left[(2x+1)^2(3x^2+4x+1) + x(x+1)^2(x^2+3x+1)\right] \\
&= \left[x^{k+2} + (2^{k+1}+k+1)x^{k+1} + \left((k+1)\cdot 2^k + \frac{k(k+1)}{2}\right)x^k + \cdots\right] \\
&\quad \cdot \left[x^{k+1} + (2^k+k)x^k + \left(k \cdot 2^{k-1} + \frac{k(k-1)}{2}\right)x^{k-1} + \cdots\right] \\
&\quad \cdot \left[x^5 + 17x^4 + 36x^3 + \cdots\right].
\end{aligned} \tag{21}$$

We can divide the first term expression into three factors $A \cdot B \cdot C$ such that:

$$\begin{aligned}
A &= \left[x^{k+2} + (2^{k+1}+k+1)x^{k+1} + \left((k+1)\cdot 2^k + \frac{k(k+1)}{2}\right)x^k + \cdots\right], \\
B &= \left[x^{k+1} + (2^k+k)x^k + \left(k \cdot 2^{k-1} + \frac{k(k-1)}{2}\right)x^{k-1} + \cdots\right], \\
C &= \left[x^5 + 17x^4 + 36x^3 + \cdots\right].
\end{aligned} \tag{22}$$

Expand the second term in that sum:

$$\begin{aligned}
x \cdot I(G - N[v_0]) &= x \cdot (2x+1)^{2k+3} \cdot (3x^2+4x+1) \\
&= x(3x^2+4x+1) \cdot \left[\sum_{i=0}^{2k+3}\binom{2k+3}{i}(2x)^i\right] \\
&= x(3x^2+4x+1) \cdot \left[(2x)^{2k+3} + \cdots\right] \\
&= 3 \cdot 2^{2k+3}x^{2k+6} + \cdots
\end{aligned} \tag{23}$$

The highest exponent that one can reach is $2k + 8$. We obtain it by taking the highest exponent from every factor, that is, from factor A we choose x^{k+2}, from factor B we choose x^{k+1}, and while from C factor we choose x^5, so

$$x^{n+2} \cdot x^{n+1} \cdot x^5 = x^{2k+8}. \tag{24}$$

with the coefficient equal to 1.

To calculate the coefficient of x^{2k+7} we have three following options:

- multiply x^{k+2} from A, x^{k+1} from B, and $17x^4$ from C,
- multiply x^{k+2} from A, $(2^k + k)x^k$ from B, and x^5 from C,
- multiply $(2^{k+1} + k + 1)x^{k+1}$ from A, x^{k+1} from B, and x^5 from C.

In such a way we obtain the following:

$$\begin{aligned} x^{k+2} \cdot x^{k+1} \cdot 17x^4 + x^{k+2} \cdot (2^k + k)x^k \cdot x^5 + (2^{k+1} + k + 1)x^{k+1} \cdot x^{k+1} \cdot x^5 \\ = (2k + 2^k + 2^{k+1} + 18)x^{2k+7}. \end{aligned} \tag{25}$$

with the coefficient equal to $2k + 2^k + 2^{k+1} + 18$.

To calculate the coefficient of x^{2k+6} we have six following options:

- multiply x^{k+2} from A, x^{k+1} from B, and $36x^3$ from C,
- multiply x^{k+2} from A, $(2^k + k)x^k$ from B, and $17x^4$ from C,
- multiply x^{k+2} from A, $\left(k \cdot 2^{k-1} + \frac{k(k-1)}{2}\right)x^{k-1}$ from B, and x^5 from C,
- multiply $(2^{k+1} + k + 1)x^{k+1}$ from A, x^{k+1} from B, and $17x^4$ from C,
- multiply $(2^{k+1} + k + 1)x^{k+1}$ from A, $(2^k + k)x^k$ from B, and x^5 from C,
- multiply $\left((k+1) \cdot 2^k + \frac{k(k+1)}{2}\right)x^k$ from A, x^{k+1} from B, and x^5 from C.

Finally, we add the coefficient $3 \cdot 2^{2k+3}$ from the second term. In such a way we obtain the following:

$$\begin{aligned} x^{k+2} \cdot x^{k+1} \cdot 36x^3 + x^{k+2} \cdot (2^k + k)x^k \cdot 17x^4 + x^{k+2} \cdot \left(k \cdot 2^{k-1} + \frac{k(k-1)}{2}\right)x^{k-1} \cdot x^5 \\ + (2^{k+1} + k + 1)x^{k+1} \cdot x^{k+1} \cdot 17x^4 + (2^{k+1} + k + 1)x^{k+1} \\ \cdot (2^k + k)x^k \cdot x^5 + \left((k+1) \cdot 2^k + \frac{k(k+1)}{2}\right)x^k \cdot x^{k+1} \cdot x^5 + 3 \cdot 2^{2k+3}x^{2k+6} \\ = \left(\frac{1}{2}k(4k + 9 \cdot 2^k + 70) + 53 \cdot 2^k + 13 \cdot 2^{2k+1} + 53\right)x^{2k+6}. \end{aligned} \tag{26}$$

with the coefficient equals to $\frac{1}{2}k\left(4k + 9 \cdot 2^k + 70\right) + 53 \cdot 2^k + 13 \cdot 2^{2k+1} + 53$. Thus the independence polynomial is:

$$I(G) = x^{2k+8} + \left(2k + 2^k + 2^{k+1} + 18\right)x^{2k+7} + \left[\frac{1}{2}k\left(4k + 9 \cdot 2^k + 70\right) + 53 \cdot 2^k + 13 \cdot 2^{2k+1} + 53\right]x^{2k+6} + \cdots \tag{27}$$

Now, let us prove the non-log-concavity of the x^{2k+7} term in the independence polynomial:

$$\left(2k + 2^k + 2^{k+1} + 18\right)^2 < 1 \cdot \left[\frac{1}{2}k\left(4k + 9 \cdot 2^k + 70\right) + 53 \cdot 2^k + 13 \cdot 2^{2k+1} + 53\right]. \tag{28}$$

The left-hand side and the right-hand side are equal to $k \approx 2.9251$ so the left-hand side is smaller than the right-hand side from $k = 3$ and above.

8. Conclusions

In this chapter, we have presented a linear algorithm that uses dynamic programming to compute independence polynomials of trees. It allows us to check all trees up to 26 vertices, to find 2 trees of order 26 with non-log-concave independence polynomials, and, finally, to construct two infinite families of trees having non-log-concave independence polynomials.

Author details

Ohr Kadrawi[1*], Vadim E. Levit[1], Ron Yosef[2] and Matan Mizrachi[2]

1 Ariel University, Ariel, Israel

2 HIT-Holon Institute of Technology, Holon, Israel

*Address all correspondence to: orka@ariel.ac.il

References

[1] Gutman I, Harary F. Generalizations of the matching polynomial. Utilitas Mathematica. 1983;**24**:97-106

[2] Hoede C, Li X. Clique polynomials and independent set polynomials of graphs. Discrete Mathematics. 1994;**125**: 219-228

[3] Chudnovsky M, Seymour P. The roots of the stable set polynomial of a clawfree graph. Journal of Combinatorial Theory, Series B. 2007;**97**:350-357. DOI: 10.1016/j.jctb.2006.06.001

[4] Ball T, Galvin D, Hyry C, Weingartner K. Independent set and matching permutations. Journal of Graph Theory. 2022;(99):40-57. DOI: 10.1002/jgt.22724

[5] Basit A, Galvin D. On the independent set sequence of a tree. The Electronic Journal of Combinatorics. 2021;**28**(3):#P3.23. DOI: 10.37236/9896

[6] Stanley RP. Log-concave and unimodal sequences in algebra, combinatorics, and geometry. Annals of the New York Academy of Sciences. 1989;**576**:500-535

[7] Brenti F. Log-concave and unimodal sequences in algebra, combinatorics, and geometry: An update, "Jerusalem Combinatorics'93". Contemporary Mathematics. 1994;**178**:71-89

[8] Brenti F. Unimodal, log-concave and Polya frequency sequances in Combinatorics. Memoris Amer, Math, Soc. 1990;**108**:729-756

[9] Bender EA, Canfield ER. Log-concavity and related properties of the cycle index polynomials. Journal of Combinatorial Theory A. 1996;**74**:57-70

[10] Boros G, Moll VH. A sequence of unimodal polynomials. Journal of Mathematical Analysis and Applications. 1999;**237**:272-287

[11] Brown JI, Colbourn CJ. On the log-concavity of reliability and matroidal sequences. Advances in Applied Mathematics. 1994;**15**:114-127

[12] Dukes WMB. On a unimodality conjecture in matroid theory. Discrete Mathematics and Theoretical Computer Science. 2002;**5**:181-190

[13] Levit VE, Mandrescu E. On unimodality of independence polynomials of some well-covered trees. Discrete Mathematics and Theoretical Computer Science. 2003;**4**:237-256

[14] Schwenk AJ. On unimodal sequences of graphical invariants. Journal of Combinatorial Theory B. 1981;**30**: 247-250

[15] Hamidoune YO. On the number of independent k-sets in a claw-free graph. Journal of Combinatorial Theory B. 1990;**50**:241-244

[16] Chudnovsky M, Seymour P. The roots of the independence polynomial of a claw-free graph. Journal of Combinatorial Theory, Series B. 2007;**97**: 350-357

[17] Beaton I, Brown JI. On the unimodality of domination polynomials. Graphs and Combinatorics. 2022;**38**:90. DOI: 10.1007/s00373-022-02487-x

[18] Horrocks DGC. The numbers of dependent k-sets in a graph are log concave. Journal of Combinatorial Theory, Series B;**84**(2002):180-185. DOI: 10.1006/jctb.2001.2077

[19] Alavi Y, Malde PJ, Schwenk AJ, Erdös P. The vertex independence sequence of a graph is not constrained. Congressus Numerantium. 1987;**58**:15-23

[20] Yosef R, Mizrachi M, Kadrawi O. On unimodality of independence polynomials of trees. 2021. Available from: https://arxiv.org/pdf/2101.06744v3.pdf. [Accessed: December 8, 2022]

[21] Radcliffe AJ. Personal communication (mentioned in Ball T, Galvin D, Hyry C, Weingartner K. Independent set and matching permutations. Journal of Graph Theory. 2022;(99):40-57.)

[22] Levit VE, Mandrescu E. Very well-covered graphs with log-concave independence polynomials. Carpathian J. Math. 2004;**20**:73-80

[23] Galvin D. REGS 2011. Available from: https://faculty.math.illinois.edu/west/regs/stasetseq.html. [Accessed: December 8, 2022]

[24] Levit VE, Mandrescu E. The independence polynomial of a graph - a survey. In: Proceedings of the 1st International Conference on Algebraic Informatics. Thessaloniki: Aristotle University of Thessaloniki; 2005. pp. 231-252

[25] Arocha JL. Propriedades del polinomio independiente de un grafo. Revista Ciencias Matematicas. 1984;**3**: 103-110

[26] Ferrin GM. Independence polynomials [Thesis]. Columbia, South Carolina: University of South Carolina; 2014

[27] Bodlaender HL. A linear time algorithm for finding tree-decompositions of small treewidth. SIAM. 1996;**25**:1305-1317

[28] Tittmann P. Graph Polynomials: The Ethernal Book. Productivity Press; 2021

Chapter 8

Applications of Orthogonal Polynomials to Subclasses of Bi-Univalent Functions

Adnan Ghazy AlAmosush

Abstract

Orthogonal polynomials have been studied extensively by Legendre in 1784. They are representatively related to typically real functions, which played an important role in the geometric function theory, and its role of estimating coefficient bounds. This chapter associates certain bi-univalent functions with certain orthogonal polynomials, such as Gegnbauer polynomials and Horadam polynomials, and then explores some properties of the subclasses in hand. This chapter is concerned with the connection between orthogonal polynomials and bi-univalent functions. Our purpose is to introduce certain classes of bi-univalent functions by means of Gegenbauer polynomials and Hordam polynomials. Bounds for the initial coefficients of $|a_2|$ and $|a_3|$, and results related to Fekete–Szegö functional are obtained.

Keywords: analytic functions, univalent and bi-univalent functions, Fekete-Szegö problem, Gegnbauer polynomials, Horadam polynomials, pseudo-starlike functions, pseudo-convex functions coefficient bounds, subordination

1. Introduction

Orthogonal polynomials appear in many areas of mathematics and play a vital role in the development of numerical and analytical approaches. Also, many mathematicians have been interested in its subjects. Orthogonal polynomials are gaining traction in current different mathematics, such as operator theory, number theory, special functions, analytic functions, and approximation theory. Also, has a wide range applications in physics and engineering fields. The subject of orthogonal polynomials finds its origins in 1784 by Legendre [1]. In the 18th century, the first examples of orthogonal polynomials were developed by brilliant mathematicians, before the general theory, which appeared in the 19th century. Orthogonal polynomials have been found to have connections with trigonometric, hypergeometric, Bessel, and elliptic functions, helping to solve certain problems in the theory of differential and integral equations, and in quantum mechanics and mathematical statistics. Up until the late 20th century, Szegö [2] covered most of the general theories along with all standard formulas for the three classical orthogonal polynomials. The connection of orthogonal polynomials with other branches of mathematics is very deep and impressive.

Officially, the classes of polynomials P_n defined over a range $[a, b]$ are orthogonal satisfy

$$degP_n = n, \quad n = 0,1,2,3, \ldots$$

and

$$\int_a^b P_n P_m W(x)dx = 0, \;\; m \neq n,$$

where $W(x)$ is nonnegative function in (a, b).

Orthogonal polynomials are among the most often studied polynomials, such as Gegenbauer polynomials, Hordam polynomials, Chebyshev polynomials of the first and the second kind, Laguerre polynomials, and Jacobi polynomial. Recently, several papers from a theoretical point of view and in the case of bi-univalent functions have been studied.

The main goal of this chapter focuses on some original results of bi-univalent functions by using Gegenbauer polynomials and Hordam polynomials. Estimates on the initial Taylor-Maclaurin coefficients and the Fekete-Szegö inequalities for some subclasses of bi univalent functions are obtained. Also, we give several illustrative examples of the bi-univalent function subclass, which we introduce here. To do so, we take into account the following definitions.

Let $\mathcal{A}$ represents the class of all functions of the form

$$f(z) = z + \sum_{n=2}^{\infty} a_n z^n, \tag{1}$$

which are analytic in the open unit open disk $U = \{z : z \in C, |z| < 1\}$, and let $\mathcal{S}$ be the class of all functions in $\mathcal{A}$, which are univalent and normalized by the conditions

$$f(0) = 0 = f'(0) - 1$$

in U.

For any two analytic function f_1 and f_2 in unit disk U, we say that f_1 is subordinate to f_2, and denoted by $f_1 \prec f_2$, if there exists Schwarz function

$$\varpi(z) = \sum_{n=1}^{\infty} \mathbf{c}_n z^n \quad (\varpi(0) = 0, |\varpi(z)| < 1), \tag{2}$$

analytic in U such that

$$f_1(z) = f_2(\varpi(z)) \qquad (z \in U), \tag{3}$$

where $|\mathbf{c}_n| \leq 1$ (see [3] for $\varpi(z)$).

In particular, it is known that

$$f(z) \prec g(z) \;\; (z \in U) \; \Leftrightarrow f(0) = g(0) \quad and \quad f(U) \subset g(U).$$

Thus, clearly, every univalent function f has an inverse f^{-1}, defined by

$$f^{-1}(f(z)) = z \quad (z \in U),$$

and

$$f^{-1}(f(w)) = w \quad \left(|w| < r_0(f); r_0(f) \geq \frac{1}{4}\right),$$

where

$$f^{-1}(w) = w + a_2 w^2 + (2a_2^2 - 3a_3)w^3 - (5a_2^3 - 5a_2 a_3 + a_4)w^4 + \cdots. \tag{4}$$

If f and f^{-1} are univalent in U, then a function $f \in \mathcal{A}$ is called *bi-univalent*.

The study of the class Σ of bi-univalent functions was discussed by Lewin [4] while Brannan and Taha [5] derived estimates for the initial coefficients. Lately, Srivastava et al. [6] have actually revived the investigation of analytic and bi-univalent functions. Several researchers have investigated and examined various subclasses of analytic and bi-univalent functions, one can refer to the works of [7–20].

Ma and Minda [21] investigated the class of starlike and convex functions as the following

$$S^*(\phi) = \left\{f : f \in \mathcal{A}, \frac{zf'(z)}{f(z)} \prec \phi(z)\right\}, \ z \in U,$$

and

$$C(\phi) = \left\{f : f \in \mathcal{A}, 1 + \frac{zf''(z)}{f'(z)} \prec \phi(z)\right\}, \ z \in U.$$

Clear that, if $f(z) \in C(\phi)$, then $zf'(z) \in S^*(\phi)$.

Initiating an investigation on properties of bi-univalent functions linked by Gegenbauer polynomials and Hordam polynomials will be discussed in the following sections.

1.1 Applications of Gegenbauer polynomials to subclasses of bi-univalent functions

This section is devoted to studying and discussing Gegenbauer Polynomials by means of bi-univalent function. We first introduce the following definitions.

A generating function of Gegenbauer polynomials of the sequence $C_n^\alpha(x), n \in \mathbb{N}$ is defined by the following:

$$H_\gamma(x, z) = \sum_{n=0}^{\infty} C_n^\alpha(x) z^n = \frac{1}{(1 - 2xz + z^2)^\alpha}, \tag{5}$$

where α is nonzero real constant, $x \in [-1, 1], z \in \mathbb{U}$, and $C_n^\alpha(x)$ is defined by

$$C_n^{\alpha}(x) = \frac{2x(n+\alpha-1)C_{n-1}^{\alpha}(x) - (n+2\alpha-2)C_{n-2}^{\alpha}(x)}{n}. \tag{6}$$

It is clear that, $C_0^{\alpha}(x) = 1, C_1^{\alpha}(x) = 2\alpha x$ and $C_2^{\alpha}(x) = 2\alpha(1+\alpha)x^2 - \alpha$. Furthermore, we present some particular cases of $C_n^{\alpha}(x)$ as follows:

1. For $\alpha = 0$, we get the Chebyshev polynomials.

2. For $\alpha = \frac{1}{2}$, we get the Legendre polynomials.

A class of starlike bi-univalent functions is linked with Gegenbauer polynomial as follows.

Definition 2.1. Let $\gamma \in (-1/2, \infty)\backslash\{0\}$, and $0 \le \lambda \le 1$, $x \in (1/2, 1)$. A function $f \in \Sigma$ given by Eq. (1) is said to be in the class $\mathcal{Q}_{\Sigma_s}(H_\gamma)$ if there exist the following functions:

$$g(z) = z + \sum_{n=2}^{\infty} b_n z^n \in \mathcal{S}^*(1/2), \qquad G(w) = w + \sum_{n=2}^{\infty} t_n w^n \in \mathcal{S}^*(1/2)$$

and the conditions are fulfilled

$$\frac{zf'(z)}{f(z)} \prec H_\gamma(x,z) \qquad (z \in \mathbb{U}), \tag{7}$$

and

$$\frac{wF'(w)}{F(w)} \prec H_\gamma(x,w) \qquad (w \in \mathbb{U}), \tag{8}$$

where F is the inverse of f is defined by Eq. (4) and H_γ is the generating function of the Gegenbauer polynomial given by Eq. (5).

Also, a class of convex bi-univalent functions linked with Gegenbauer polynomial as follows:

Definition 2.2. Let $\gamma \in (-1/2, \infty)\backslash\{0\}$ and $0 \le \lambda \le 1$, $x \in (1/2, 1)$. A function $f \in \Sigma$ given by (??) is said to be in the class $\mathcal{Q}^*_{\Sigma_s}(H_\gamma)$ if there exist the following functions:

$$g(z) = z + \sum_{n=2}^{\infty} b_n z^n \in \mathcal{S}^*(1/2), \qquad G(w) = w + \sum_{n=2}^{\infty} t_n w^n \in \mathcal{S}^*(1/2)$$

and the conditions are fulfilled

$$1 + \frac{zf''(z)}{f'(z)} \prec H_\gamma(x,z) \qquad (z \in \mathbb{U}), \tag{9}$$

and

$$\frac{wF''(w)}{F'(w)} \prec H_\gamma(x,w) \qquad (w \in \mathbb{U}), \tag{10}$$

where F is the inverse of f is defined by Eq. (4) and H_γ is the generating function of the Gegenbauer polynomial given by Eq. (5).

Next section, we discuss some recent results of bi-univalent functions for several families associated with Gegenbauer polynomials and Hordam polynomials, respectively. For the proofs and details of the main theorems, one can refer to [7–9], respectively.

1.2 Initial Taylor coefficient estimates for the functions of $\mathcal{Q}_{\Sigma_s}(H_\gamma)$ and $\mathcal{Q}^*_{\Sigma_s}(H_\gamma)$

We begin this section by defining Fekete-Szegö inequality, which was given by Fekete and Szegö [2] and defined as follows.

If $f \in S$ and η is real, then

$$|a_3 - \eta a_2^2| \leq 1 + 2e^{\frac{-2\eta}{1-\mu}}. \tag{11}$$

This bound is sharp.

Theorem 2.1. Let the function f given by Eq. (1) be in the class $\mathcal{Q}_{\Sigma_s}(H_\gamma)$. Then

$$|a_2| \leq \frac{2|\gamma|x\sqrt{2|\gamma|x}}{\sqrt{|4\gamma(\gamma-1)x^2 + 2\gamma|}} \tag{12}$$

and

$$|a_3| \leq \gamma^2 x^2 + \frac{|\gamma|x}{3}. \tag{13}$$

For some $\eta \in \mathbb{R}$, we have

$$|a_3 - \eta a_2^2| \leq \begin{cases} \frac{|\gamma|x}{3}, & if \quad |\eta - 1| \leq \left|\frac{1-2x^2}{6\gamma x^2}\right| \\ \frac{2\gamma^2 x^3 |1-\eta|}{|1-2x^2|}, & if \quad |\eta - 1| \leq \left|\frac{1-2x^2}{6\gamma x^2}\right| \end{cases} \tag{14}$$

For $\gamma = 1$ in Theorem 2.1, we have the following corollary.

Corollary 2.1. Let the function f given by Eq. (1) be in the class $\mathcal{Q}_{\Sigma_s}(H)$. Then

$$|a_2| \leq 2x\sqrt{x} \tag{15}$$

and

$$|a_3| \leq x^2 + \frac{x}{3}. \tag{16}$$

For some $\eta \in \mathbb{R}$, we have

$$|a_3 - \eta a_2^2| \leq \begin{cases} \frac{x}{3}, & if \quad |\eta - 1| \leq \left|\frac{1-2x^2}{6x^2}\right| \\ \frac{2x^3|1-\eta|}{|1-2x^2|}, & if \quad |\eta - 1| \leq \left|\frac{1-2x^2}{6x^2}\right| \end{cases} \tag{17}$$

Theorem 2.2. Let the function f given by Eq. (1) be in the class $\mathcal{Q}^{*}_{\Sigma_s}(H_\gamma)$. Then

$$|a_2| \leq \frac{2|\gamma|x\sqrt{2|\gamma|x}}{\sqrt{|2\gamma(\gamma-1)x^2+\gamma|}} \tag{18}$$

and

$$|a_3| \leq 4\gamma^2x^2 + |\gamma|x. \tag{19}$$

For some $\eta \in \mathbb{R}$, we have

$$|a_3 - \eta a_2^2| \leq \begin{cases} |\gamma|x, & if \quad |\eta - 1| \leq \left|\frac{2\gamma x^2 - 2x^2 + 1}{2\gamma x^2}\right| \\ \frac{8|\gamma|^3x^3|1-\eta|}{|2\gamma(\gamma-1)x^2+\gamma|}, & if \quad |\eta - 1| \leq \left|\frac{2\gamma x^2 - 2x^2 + 1}{2\gamma x^2}\right| \end{cases} \tag{20}$$

For $\gamma = 1$ in Theorem 2.2, we have the following corollary.
Corollary 2.2. Let the function f given by Eq. (1) be in the class $\mathcal{Q}^{*}_{\Sigma_s}(H_\gamma)$. Then

$$|a_2| \leq 2x\sqrt{2x} \tag{21}$$

and

$$|a_3| \leq 4x^2 + x. \tag{22}$$

For some $\eta \in \mathbb{R}$, we have

$$|a_3 - \eta a_2^2| \leq \begin{cases} x, & if \quad |\eta - 1| \leq \frac{1}{2x^2} \\ 8x^3|1-\eta|, & if \quad |\eta - 1| \leq \frac{1}{2x^2} \end{cases} \tag{23}$$

Next section is devoted to studying and discussing Hordam polynomials by means of bi-univalent function.

1.3 Applications of Hordam polynomials to subclasses of bi-univalent functions

We begin this section by introducing the recurrence relation of the Hordam polynomials, which was studied by Horzum and Koçer [22] as follows.

$$h_n(x) = pxh_{n-1}(x) + qh_{n-2}(x); \quad (n \in \mathbb{N} \geq 2), \tag{24}$$

with

$$h_1(x) = a, \quad h_2(x) = bx, \quad h_3(x) = pbx^2 + pq, \tag{25}$$

where $a, b, p,$ and q are some real constants, and the characteristic equation of above recurrence relation is

$$t^2 - pxt - q = 0, \tag{26}$$

with two real roots:

$$\alpha = \frac{px + \sqrt{p^2x^2 + 4q}}{2},$$

and

$$\beta = \frac{px - \sqrt{p^2x^2 + 4q}}{2}.$$

Selecting particular values of $a, b, p,$ and q reduces to special various polynomials as follows:

- If $a = b = p = q = 1$, the Fibonacci polynomials sequence is obtained

$$F_n(x) = xF_{n-1}(x) + F_{n-2}(x), \quad F_1(x) = 1, \quad F_2(x) = x.$$

- If $a = 2, b = p = q = 1$, the Lucas polynomials sequence is acquired

$$L_{n-1}(x) = xL_{n-2}(x) + L_{n-3}(x), \quad L_0(x) = 2, \quad L_1(x) = x.$$

- If $a = q = 1, b = p = 2$, the Pell polynomials sequence is attained

$$P_n(x) = 2xP_{n-1}(x) + P_{n-2}(x), \quad p_1(x) = 1, \quad P_2(x) = 2x.$$

- If $a = b = p = 2, q = 1$, the Pell-Lucas polynomials sequence is obtained

$$Q_{n-1}(x) = 2xQ_{n-2}(x) + Q_{n-3}(x), \quad Q_0(x) = 2, \quad Q_1(x) = 2x.$$

- If $a = 1, b = p = 2, q = -1$, the Chebyshev polynomials of second kind sequence are acquired

$$U_{n-1}(x) = 2xU_{n-2}(x) + U_{n-3}(x), \quad U_0(x) = 1, \quad U_1(x) = 2x.$$

- If $a = b = 1, p = 2, q = -1$, the Chebyshev polynomials of first kind sequence are obtained

$$T_{n-1}(x) = 2xT_{n-2}(x) + T_{n-3}(x), \quad T_0(x) = 1, \quad T_1(x) = x.$$

- If $x = 1$, the Horadam numbers sequence is derived

$$h_{n-1}(1) = ph_{n-2}(1) + qh_{n-3}(1), \quad h_0(1) = a, \quad h_1(1) = b.$$

For more details related to these polynomial sequences succession, can refer to [22–25].

The generating function of the Horadam polynomials $h_n(x)$ is studied by Horadam [23] and defined as follows:

$$\Omega(x,z)=\frac{a+(b-ap)xt}{1-pxt-qt^2}=\sum_{n=1}^{\infty}h_n(x)z^{n-1}. \tag{27}$$

New subclasses of the bi-univalent function class Σ associated with Horadam polynomial are presented in the following.

Definition 3.1. A function $f\in\Sigma'$ given by Eq. (1) is said to be in the class $\Sigma'(x)$ if the following subordinations hold:

$$f'(z)\prec\Omega(x,z)+1-\alpha \tag{28}$$

and

$$g'(w)\prec\Omega(x,w)+1-\alpha, \tag{29}$$

where the real constants a, b, and q are as in Eq. (25) and $g=f^{-1}$ is given by Eq. (4).

Theorem 3.1. Let the function $f\in\Sigma'$ given by Eq. (1) be in the class $\Sigma'(x)$. Then

$$|a_2|\le\frac{|bx|\sqrt{|bx|}}{\sqrt{\left|bx^2(3b-4p)-4aq\right|}} \tag{30}$$

$$|a_3|\le\frac{|bx|}{3}+\frac{(bx)^2}{4}, \tag{31}$$

and for some $\eta\in R$,

$$|a_3-\eta a_2^2|\le\left\{\begin{array}{ll}\dfrac{|2bx|}{3} & ,\ |\eta-1|\le 1-\dfrac{|4(pbx^2+qa)|}{3b^2x^2}\\ \dfrac{|bx|^3|1-\eta|}{3b^2x^2-4(pbx^2+qa)} & ,\ |\eta-1|\ge 1-\dfrac{|4(pbx^2+qa)|}{3b^2x^2}\end{array}\right\}. \tag{32}$$

In light of relation (27), Theorem 3.1, we can readily deduce the following corollaries.

Corollary 3.1. For $t\in(1/2,1)$, let the function $f\in\Sigma'$ given by Eq. (1) be in the class $\Sigma'(t)$. Then

$$|a_2|\le\frac{t\sqrt{2t}}{\sqrt{|1-t^2|}} \tag{33}$$

$$|a_3|\le\frac{2t}{3}+t^2, \tag{34}$$

and for some $\eta\in R$,

$$|a_3-\eta a_2^2|\le\left\{\begin{array}{ll}\dfrac{4t}{3} & ,\ |\eta-1|\le\dfrac{1-t^2}{3t^2}\\ \dfrac{2|\eta-1|}{1-t^2} & ,\ |\eta-1|\ge\dfrac{1-t^2}{3t^2}\end{array}\right\}. \tag{35}$$

Taking $\eta = 1$ in Corollary 3.1, we get the following corollary.

Corollary 3.2. For $t \in (1/2, 1)$, let the function $f \in \Sigma'$ given by Eq. (1) be in the class $\Sigma'(t)$. Then

$$|a_3 - a_2^2| \leq \frac{4t}{3}. \tag{36}$$

The class S_s^* of functions starlike with respect to symmetric points is introduced by Sakaguchi [26], consisting of functions $f \in S$ that satisfy the following condition

$$\Re\left(\frac{2zf'(z)}{f(z) - f(-z)}\right) > 0, \qquad z \in \mathbb{U}.$$

Similarly, the class K_s of functions convex with respect to symmetric points is introduced by Wang et al. [27], consisting of functions $f \in S$ that satisfy the following condition

$$\Re\left(\frac{2(zf'(z))'}{f'(z) + f'(-z)}\right) > 0, \qquad z \in \mathbb{U}.$$

Moreover, Ravichandran [28] introduced the following two subclasses:

For such a function ϕ, then a function $f \in A$ is in the class $S_s^*(\phi)$ if

$$\frac{2zf'(z)}{f(z) - f(-z)} \prec \phi(z), \quad z \in \mathbb{U},$$

and in the class $K_s(\phi)$ if

$$\frac{2(zf'(z))'}{f'(z) + f'(-z)} \prec \phi(z) \quad z \in \mathbb{U}.$$

Recently, a new class L_λ of λ-pseudo-starlike functions is defined by Babalola [29] as the following:

Let $f \in A$ and $\lambda \geq 1$ is real. Then $f(z) \in L_\lambda$ of λ-pseudo-starlike functions in $\mathbb{U}$ if and only if

$$\Re\left\{\frac{z(f'(z))^\lambda}{f(z)}\right\} \geq 0, \quad z \in \mathbb{U}.$$

More recently, the author introduced and studied two subclasses $L\Sigma(\lambda, \alpha, x)$ and $M\Sigma(\lambda, \alpha, x)$ of λ-pseudo-bi-univalent functions with respect to symmetrical points linked by the Horadam polynomials $h_n(x)$ and the generating function $\Omega(x,z)$ as follows.

Definition 3.2. A function $f \in \Sigma$ given by Eq. (1) is said to be in the class $L\Sigma(\lambda, \alpha, x)$ if the following conditions are satisfied:

$$(1-\alpha)\frac{2z[f'(z)]^\lambda}{f(z) - f(-z)} + \alpha\frac{2\left[(zf'(z))'\right]^\lambda}{[f(z) - f(-z)]'} \prec \Omega(x,z) + 1 - \alpha \tag{37}$$

and

$$(1-\alpha)\frac{2w[g'(w)]^{\lambda}}{g(w)-g(-w)}+\alpha\frac{2\left[(wg'(w))'\right]^{\lambda}}{[g(w)-g(-w)]'}\prec\Omega(x,w)+1-\alpha \tag{38}$$

where the real constants a, b, and q are as in Eq. (25) and $g(w)=f^{-1}(z)$ is given by Eq. (4).

Definition 3.3. A function $f\in\Sigma$ given by Eq. (1) is said to be in the class $M\Sigma(\lambda,\alpha,x)$ if the following conditions are satisfied:

$$\left(\frac{2z\left[f'(z)\right]^{\lambda}}{f(z)-f(-z)}\right)^{\alpha}\left(\frac{2\left[\left(zf'(z)\right)'\right]^{\lambda}}{[f(z)-f(-z)]'}\right)^{1-\alpha}\prec\Omega(x,z)+1-\alpha \tag{39}$$

and

$$\left(\frac{2w[g'(w)]^{\lambda}}{g(w)-g(-w)}\right)^{\alpha}\left(\frac{2\left[(wg'(w))'\right]^{\lambda}}{[g(w)-g(-w)]'}\right)^{1-\alpha}\prec\Omega(x,w)+1-\alpha \tag{40}$$

where the real constants a, b, and q are as in Eq. (25) and $g(w)=f^{-1}(z)$ is given by Eq. (4).

In the next theorems, the coefficient bounds and Fekete-Szegö type inequalities for the function subclasses $L\Sigma(\lambda,\alpha,x)$ and $M\Sigma(\lambda,\alpha,x)$, respectively.

Theorem 3.2. Let the function $f\in\Sigma$ given by Eq. (1) be in the class $L\Sigma(\lambda,\alpha,x)$. Then

$$|a_2|\le\frac{|bx|\sqrt{|bx|}}{\sqrt{\left|\left[\left((2\lambda^2+\lambda-1)+2\alpha(3\lambda^2-1)\right)b-4p\lambda^2(1+\alpha)^2\right]bx^2-4qa\lambda^2(1+\alpha)^2\right|}} \tag{41}$$

$$|a_3|\le\frac{|bx|}{(3\lambda-1)(1+2\alpha)}+\frac{(bx)^2}{4\lambda^2(1+\alpha)^2}, \tag{42}$$

and for some $\eta\in\mathbb{R}$,

$$|a_3-\eta a_2^2|\le\begin{cases}\dfrac{|bx|}{(3\lambda-1)(1+2\alpha)}, & |\eta-1|\le\dfrac{1}{2(3\lambda-1)(1+2\alpha)}|A|\\[2ex] \dfrac{|bx|^3|1-\eta|}{\left|\left[(2\lambda^2+\lambda-1)+2\alpha(3\lambda^2-1)\right][bx]^2-4\lambda^2(1+\alpha)^2(pbx^2+qa)\right|}, & |\eta-1|\ge\dfrac{1}{2(3\lambda-1)(1+2\alpha)}|A|\end{cases} \tag{43}$$

where $A=\left(2\lambda^2+\lambda-1\right)+2\alpha\left(3\lambda^2-1\right)-\frac{4\lambda^2(1+\alpha)^2\left(pbx^2+qa\right)}{b^2x^2}$.

For $\alpha=0$ in Theorem 3.2, we have the following result.

Corollary 3.3. Let the function $f\in\Sigma$ given by Eq. (1) be in the class $L\Sigma(\lambda,x)$. Then

$$|a_2|\le\frac{|bx|\sqrt{|bx|}}{\sqrt{\left|\left[(2\lambda^2+\lambda-1)b-4p\lambda^2\right]bx^2-4qa\lambda^2\right|}} \tag{44}$$

$$|a_3| \le \frac{|bx|}{3\lambda - 1} + \frac{(bx)^2}{4\lambda^2}, \tag{45}$$

and for some $\eta \in \mathbb{R}$,

$$|a_3 - \eta a_2^2| \le \begin{cases} \dfrac{|bx|}{3\lambda - 1} & , \ |\eta - 1| \le \dfrac{1}{2(3\lambda - 1)}|A_0| \\ \dfrac{|bx|^3 |1-\eta|}{|(2\lambda^2 + \lambda - 1)[bx]^2 - 4\lambda^2(pbx^2 + qa)|} & , \ |\eta - 1| \ge \dfrac{1}{2(3\lambda - 1)}|A_0| \end{cases} \tag{46}$$

where $A_0 = (2\lambda^2 + \lambda - 1) - \frac{4\lambda^2(pbx^2 + qa)}{b^2x^2}$.

For $\alpha = 1$ Theorem 3.2, we have the following result.

corollary 3.3.

Let the function $f \in \Sigma$ given by Eq. (1) be in the class $L\Sigma(\lambda, 1, x)$. Then

$$|a_2| \le \frac{|bx|\sqrt{|bx|}}{\sqrt{|[((2\lambda^2 + \lambda - 1) + 2(3\lambda^2 - 1))b - 16p\lambda^2]bx^2 - 16qa\lambda^2|}} \tag{47}$$

$$|a_3| \le \frac{|bx|}{3(3\lambda - 1)} + \frac{(bx)^2}{16\lambda^2}, \tag{48}$$

and for some $\eta \in \mathbb{R}$,

$$|a_3 - \eta a_2^2| \le \begin{cases} \dfrac{|bx|}{3(3\lambda - 1)} & , \ |\eta - 1| \le \dfrac{1}{6(3\lambda - 1)}|A_1| \\ \dfrac{|bx|^3 |1-\eta|}{|[(2\lambda^2 + \lambda - 1) + 2(3\lambda^2 - 1)][bx]^2 - 16\lambda^2(pbx^2 + qa)|} & , \ |\eta - 1| \ge \dfrac{1}{6(3\lambda - 1)}|A_1| \end{cases} \tag{49}$$

where $A_1 = (2\lambda^2 + \lambda - 1) + 2(3\lambda^2 - 1) - \frac{16\lambda^2(pbx^2 + qa)}{b^2x^2}$.

For $\lambda = 1$ Theorem 3.2, we have the following result.

Corollary 3.4. Let the function $f \in \Sigma$ given by Eq. (1) be in the class $L\Sigma(1, \alpha, x)$. Then

$$|a_2| \le \frac{|bx|\sqrt{|bx|}}{\sqrt{\left|\left[2(1 + 2\alpha)b - 4p(1+\alpha)^2\right]bx^2 - 4qa(1+\alpha)^2\right|}} \tag{50}$$

$$|a_3| \le \frac{|bx|}{2(1 + 2\alpha)} + \frac{(bx)^2}{4(1+\alpha)^2}, \tag{51}$$

and for some $\eta \in \mathbb{R}$,

$$|a_3 - \eta a_2^2| \le \begin{cases} \dfrac{|bx|}{2(1 + 2\alpha)} & , \ |\eta - 1| \le \dfrac{1}{4(1 + 2\alpha)}|B| \\ \dfrac{|bx|^3 |1-\eta|}{|2(1+\alpha)[bx]^2 - 4(1+\alpha)^2(pbx^2 + qa)|} & , \ |\eta - 1| \ge \dfrac{1}{4(1 + 2\alpha)}|B| \end{cases} \tag{52}$$

where $B = 2(1+2\alpha) - \frac{4(1+\alpha)^2\left(pbx^2+qa\right)}{b^2x^2}$..

A function $f \in \Sigma$ given by (1.1) is said to be in the class $M\Sigma(\lambda, \alpha, x)$ if the following conditions are satisfied:

$$\left(\frac{2z\left[f'(z)\right]^{\lambda}}{f(z)-f(-z)}\right)^{\alpha}\left(\frac{2\left[\left(zf'(z)\right)'\right]^{\lambda}}{\left[f(z)-f(-z)\right]'}\right)^{1-\alpha} \prec \Omega(x,z)+1-\alpha \tag{53}$$

and

$$\left(\frac{2w\left[g'(w)\right]^{\lambda}}{g(w)-g(-w)}\right)^{\alpha}\left(\frac{2\left[\left(wg'(w)\right)'\right]^{\lambda}}{\left[g(w)-g(-w)\right]'}\right)^{1-\alpha} \prec \Omega(x,w)+1-\alpha \tag{54}$$

where the real constants a, b, and q are as in Eq. (25) and $g(w) = f^{-1}(z)$ is given by Eq. (4).

Theorem 3.3. Let the function $f \in \Sigma$ given by Eq. (1) be in the class $M\Sigma(\lambda, \alpha, x)$. Then

$$|a_2| \leq \frac{|bx|\sqrt{|bx|}}{\sqrt{\left|\left[\left(2\lambda^2(\alpha-2)^2+(\lambda+2\alpha-3)\right)b-4p\lambda^2(\alpha-2)^2\right]bx^2-4qa\lambda^2(\alpha-2)^2\right)\right|}} \tag{55}$$

$$|a_3| \leq \frac{|bx|}{(3\lambda-1)(|3-2\alpha|)}+\frac{(bx)^2}{4\lambda^2(\alpha-2)^2}, \tag{56}$$

and for some $\eta \in \mathbb{R}$,

$$|a_3-\eta a_2^2| \leq \begin{cases} \dfrac{|bx|}{(3\lambda-1)(3-2\alpha)} & , \ |\eta-1| \leq \dfrac{1}{2(3\lambda-1)(3-2\alpha)}|C| \\ \dfrac{|bx|^3|1-\eta|}{\left|\left[2\lambda^2(\alpha-2)^2+(\lambda+2\alpha-3)\right][bx]^2-4\lambda^2(\alpha-2)^2\left(pbx^2+qa\right)\right|} & , \ |\eta-1| \geq \dfrac{1}{2(3\lambda-1)(3-2\alpha)}|C| \end{cases} \tag{57}$$

where $C = 2\lambda^2(\alpha-2)^2+(\lambda+2\alpha-3) - \frac{4\lambda^2(\alpha-2)^2\left(pbx^2+qa\right)}{b^2x^2}$.

For $\lambda = 1$ Theorem 3.3, we have the following result.

Corollary 3.5. Let the function $f \in \Sigma$ given by Eq. (1) be in the class $M\Sigma(1, \alpha, x)$. Then

$$|a_2| \leq \frac{|bx|\sqrt{|bx|}}{\sqrt{\left|\left[\left(2(\alpha-2)^2+2(\alpha-1)\right)b-4p(\alpha-2)^2\right]bx^2-4qa(\alpha-2)^2\right)\right|}} \tag{58}$$

$$|a_3| \leq \frac{|bx|}{2(|3-2\alpha|)}+\frac{(bx)^2}{4(\alpha-2)^2}, \tag{59}$$

and for some $\eta \in \mathbb{R}$,

$$|a_3 - \eta a_2^2| \leq \begin{cases} \dfrac{|bx|}{2(3-2\alpha)} & , \ |\eta - 1| \leq \dfrac{1}{4(3-2\alpha)}|C_1| \\ \dfrac{|bx|^3 |1-\eta|}{\left|\left[2(\alpha-2)^2 + 2(\alpha-1)\right][bx]^2 - 4(\alpha-2)^2(pbx^2+qa)\right|} & , \ |\eta - 1| \geq \dfrac{1}{4(3-2\alpha)}|C_1| \end{cases} \tag{60}$$

where $C_1 = 2(\alpha-2)^2 + 2(\alpha-1) - \frac{4(\alpha-2)^2(pbx^2+qa)}{b^2x^2}$.

2. Conclusions

The study conducted in this chapter involves certain subclasses of bi-univalent functions examined by using Gegenbauer polynomials and Hordam polynomials, respectively, in the open unit disc. The main results are contained in which coefficient estimates are obtained for each subclass of bi-univalent functions, which could inspire researchers to focus on other aspects such as certain families of bi-univalent functions using other orthogonal polynomials.

Acknowledgements

It is a great pleasure for me to appreciate Prof. Faruk Özge and academic editor for their precious time in reviewing this chapter and putting it in their book project.

Author details

Adnan Ghazy AlAmosush
Faculty of Science, Taibah University, Saudi Arabia

*Address all correspondence to: agalamoush@taibahu.edu.sa

References

[1] Legendre A. Recherches sur lattraction des spheroides homogenes, Memoires presentes par presents par divers sa vents al. Vol. 10. Paris: Academie des Sciences de lInstitut de France; 1785. pp. 411-434

[2] Fekete M, Szegö G. Eine Bemerkung Ã41 ber ungerade schlichte Funktionen. Journal of the London Mathematical Societ. 1933;**1**(2):85-89

[3] Duren P. Univalent functions, Grundlehren der Mathematischen Wissenschaften 259. New York: Springer-Verlag; 1983

[4] Lewin M. On a coefficient problem for bi-univalent functions. Proceeding of the Amarican Mathematical Society. 1967;**18**(1):63-68

[5] Brannan DA, Taha TS. On some classes of bi-unvalent functions. In: Mazhar SM, Hamoui A, Faour NS, editors. Mathematical Analysis and its Applications, Kuwait; 1985. Vol. 3. Oxford: Pergamon Press (Elsevier Science Limited); 1988. pp. 53-60, KFAS Proceedings Series, see also Studia Universitatis Babeş-Bolyai Mathematica. 1986. pp. 70-77

[6] Srivastava HM, Mishra AK, Gochhayat P. Certain subclasses of analytic and bi-univalent function. Applied Mathematics Letters. 2010;**23**: 1188-1192

[7] Amourah A, Alamoush A, Al-Kaseasbeh M. Gegenbauer polynomials and Bi-univalent functions. Palestine Journal of Mathematics. 2021; **10**(2):625-632

[8] Alamoush A. Coefficient estimates for a new subclasses of lambda-pseudo bi-univalent functions with respect to symmetrical points associated with the Horadam polynomials. Turkish Journal of Mathematics. 2019;**3**:2865-2875

[9] Alamoush A. On a subclass of bi-univalent functions associated to Horadam polynomials. International Journal of Open Problems Complex Analysis. 2020;**12**(1):58-66

[10] Alamoush A. Certain subclasses of bi-univalent functions involving the Poisson distribution associated with Horadam polynomials. Malaya Journal of Matematik. 2019;**7**:618-624

[11] Alamoush A. Coefficient estimates for certain subclass of bi-Bazilevic functions associated with Chebyshev polynomials. Acta Universitatis Apulensis. 2019;**60**:53-59

[12] Alamoush A. On subclass of analytic bi-close-to-convex functions. International Journal of Open Problems Complex Analysis. 2021;**13**(1):10-18

[13] AlAmoush AG, Wanas AK. Coefficient estimates for a new subclass of bi-close-to-convex functions associated with the Horadam polynomials. International Journal of Open Problems Complex Analysis. 2022;**14**(1):16-26

[14] AlAmoush AG, Bulut S. On subclass of bi-close-to-convex functions by mean of the Gegenbauer polynomials. Mathématiques Appliquées. 2021;**10**: 93-101

[15] Alamoush A, Darus M. Coefficient bounds for new subclasses of bi-univalent functions using Hadamard product. Acta Universitatis Apulensi. 2014;**38**:153-161

[16] Alamoush A, Darus M. Coefficients estimates for bi-univalent of fox-wright

functions. Far East. Journal of Mathematical Sciences. 2014;**89**(2): 249-262

[17] Alamoush A, Darus M. On coefficient estimates for new generalized subclasses of bi-univalent functions. AIP Conference Proceedings. 2014;**1614**: 844-805

[18] Srivastava H, Wanas A, Murugusundaramoorthy G. A certain family of bi-univalent functions associated with the Pascal distribution series based upon the Horadam polynomials. Surveys in Mathematics and its Applications. 2021;**16**:193-205

[19] Wanas A. Horadam polynomials for a new family of λ-pseudo bi-univalent functions associated with Sakaguchi type functions. Afrika Matematika. 2021;**32**: 879-889. DOI: 10.1007/s13370-020-00867-1

[20] Altinkaya S, Yalçin S. Coefficient estimates for two new subclasses of bi univalent functions with respect to symmetric points. Journal of Function Spaces. 2015;**2015**:145242, 5 pages

[21] Ma WC, Minda D. A unified treatment of some special classes of univalent functions. In: Proceedings of the Conference on Complex Analysis. Nankai Institute of Mathematics. New York: International Press; 1992. pp. 157-169

[22] Horcum T, Kocer EG. On some properties of Horadam polynomials. International Mathematical Forum. 2009;**4**(25):1243-1252

[23] Horadam AF. Jacobsthal representation polynomials. The Fibonacci Quarterly. 1997;**35**(2):137-148

[24] Horadam AF, Mahon JM. Pell and Pell-Lucas polynomials. The Fibonacci Quarterly. 1985;**23**(1):7-20

[25] Koshy T. Fibonacci and Lucas Numbers with Applications. A Wiley-Interscience Publication, John Wiley & Sons, Inc.; 2001

[26] Sakaguchi K. On a certain univalent mapping. The Journal of the Mathematical Society of Japan. 1959;**11**: 72-75

[27] Wang GCY, Yuan SM. On certain subclasses of close-to-convex and quasi-convex functions with respectto $k-$symmetric points. Journal of Mathematical Analysis and Applications. 2006;**322**:97-106

[28] Ravichandran V. Starlike and convex functions with respect to conjugate points. Acta Mathematica. Academiae Paedagogicae Nyiregyhaziensis. 2004; **20**:31-37

[29] Babalola K. On λ-pseudo-starlike functions. Journal of Classical Analysis. 2013;**3**(2):137-147

Chapter 9

Properties of the Poly-Changhee Polynomials and Its Applications

Mohd. Saif and Raghib Nadeem

Abstract

The introduction of Changhee polynomials and numbers by Kim et al. in 2013, as published in Advances in Studies in Theoretical Physics, Volume 7, 2013, No. 20, pages 993–1003, sparked interest in their properties and identities. Building on Kim's work, the authors of this chapter define the generating functions for poly-Changhee polynomials and numbers, as well as higher-order poly-Changhee polynomials and numbers, using an umbral calculus approach. They demonstrate how these polynomials and numbers are closely related to other polynomials, such as Stirling polynomials of the first and second kind, and Daehee polynomials and numbers, by utilizing poly-logarithmic functions. The authors also derive new explicit formulas and identities for these polynomials and numbers in their research.

Keywords: Changhee polynomials, degenearate poly-changhee polynomials, generating function, higher-order changhee polynomials, poly-Bernoulli numbers

1. Introduction

In recent years, there has been significant research on Changhee polynomials and their properties, generalizations, and applications in various fields. Here is a brief overview of some of the notable works in this area:

- Kim et al. [1]: Introduced Changhee polynomials and numbers using an umbral calculus approach and obtained interesting identities and properties of these polynomials.

- Kim et al. [2]: Considered Witt-type formulas for Changhee numbers and polynomials, which are a type of recurrence relation satisfied by these numbers and polynomials.

- Kim and Kim [3]: Introduced higher-order Changhee polynomials and established relations between higher-order Changhee polynomials and special polynomials.

- Rim et al. [4]: Considered Witt-type formulas for n-th twisted Changhee numbers and polynomials, which are a generalization of Changhee numbers and polynomials with a twist parameter n.

 IntechOpen

- Jang et al. [5]: Investigated higher-order twisted Changhee polynomials and numbers, and discussed computations of zeros of these polynomials.
- Kim et al. [6, 7]: Established nonlinear differential equations satisfied by Changhee polynomials, which turned out to be useful for studying special polynomials and mathematical physics.

Moreover, Changhee polynomials and numbers have found applications in various fields, including mathematics, mathematical physics, computer science, engineering sciences, and real-world problems [1–40]. These polynomials and numbers have been used to solve problems in diverse areas, showcasing their versatility and importance in different disciplines.

Additionally, the notation $\mathbb{N}$ denotes the set of natural numbers, and $\mathbb{N}_0 = \mathbb{N} \cup \{0\}$ denotes the set of non-negative integers.

For $t \in \mathbb{C}_p$ with $|t|_p < p^{-\frac{1}{p-1}}$, the Changhee polynomials are defined by the generating function to be

$$\frac{2}{t+2}(1+t)^x = \sum_{n=0}^{\infty} Ch_n(x)\frac{t^n}{n!} \qquad \text{(see [20,23,25,30]).} \tag{1}$$

For $x = 0, Ch_n = Ch_n(0) = \frac{n!}{2^n}$ are called Changhee numbers [1]. The first few Changhee numbers Ch_n are $Ch_0 = 1, Ch_1 = \frac{1}{2}, Ch_2 = \frac{1}{2}, Ch_3 = \frac{3}{4}, Ch_4 = \frac{3}{2}, Ch_5 = \frac{15}{4}, \ldots$.

From (1), we have

$$Ch_n(x) = \sum_{l=0}^{n} \binom{n}{l} (x)_{n-l}\ Ch_l, \tag{2}$$

where $(x)_n$ is a falling factorial (also called the Pochhammer symbol and related to the gamma function in such a way that $\frac{\Gamma(x+n)}{\Gamma(x)}$, for detail (see [11, 17, 26, 41]).

$$(x)_n = x(x-1)(x-2)\cdots(x-n+1) \qquad (n \in \mathbb{N} \cup \{0\}). \tag{3}$$

Further, to find the remarkable results in the following section, we need to define classical Bernoulli $B_n(x)$ [16, 21, 33], Bernoulli polynomials of the second kind $b_n(x)$ [14], modified poly-Bernoulli numbers $C_n^{(k)}$ [32], poly-Daehee numbers $D_n^{(k)}(x)$ [15] by means of the generating functions, so before defining all these definitions, first we need to define generating function in a nutshell.

2. Generating functions

The generating function is a way of encoding an infinite sequence of numbers $\langle a_0, a_1, a_2 \cdots \rangle$ by treating them as a coefficient of a formal power series. Then, the series is called the generating function of the series. For example, if $\langle a \rangle = \langle a_0, a_1, a_2 \cdots \rangle$ be a sequence of terms, then generating function (in short we call it GF) of the above sequence is an infinite series $G(x) = \sum_{n=0}^{\infty} a_n x^n$.

- If $\langle a_n \rangle = c\ (c \neq 0)$, then GF becomes $G(x) = \frac{c}{1-x}$.

- If $\langle a_n \rangle = \begin{cases} \binom{n}{k}, & k \le n \\ 0, & k > n \end{cases}$, then GF becomes $G(x) = (1+x)^n$.
- If $\langle a_n \rangle = c \ (c \neq 0)$, then GF becomes $G(x) = \frac{c}{1-x}$.
- If $\langle a_n \rangle = \frac{n!}{2^n}$, then GF becomes $G(x) = \frac{2}{2+x} = Ch_n(0)$, the Changhee numbers described in first section. Similarly, if we choose $\langle a_n \rangle = \left(\frac{1}{2}\right)^n$, then we also get Changhee numbers. So, here we have seen that the generating function has an elementary form. In fact, more simple than the sequence itself, this is the first magic of the generating function; in many natural instances, the generating function turns out to be very simple.

There are various generating functions, such as ordinary, exponential, Bell series, etc. Similarly, we can find generating function in x and t, denoted by $G(x,t)$ and defined by the formal power series as

$$G(x,t) = \sum_{n=0}^{\infty} g_n(x) t^n, \tag{4}$$

and we say that $G(x,t)$ has generted the set $g_n(x)$. We can extend the above definition slightly by the form

$$G(x,t) = \sum_{n=0}^{\infty} c_n g_n(x) t^n, \tag{5}$$

where c_n be sequence independent of x and t. Certain properties of the polynomial set $g_n(x)$ are readily deduced from the known properties of $G(x,t)$. In this decades, many of researchers found the generating function of various types of polynomial and also generalize the idea to get new set of polynomial and numbers. In this regard, many authors can visit the references. Also, generating function plays a large role in our study of the polynomial sets. For example,

- The Legender polynomials (denoted by $P_n(x)$) is defined by the GF

$$G(x,t) = \frac{1}{\sqrt{(1 - 2xt + t^2)}} = \sum_{n=0}^{\infty} P_n(x) t^n.$$

- The Hermite polynomial (denoted by $H_n(x)$) is defined by the GF

$$G(x,t) = exp\left(2xt - t^2\right) = \sum_{n=0}^{\infty} H_n(x) t^n.$$

Note 2.1. *Let $\mathcal{A}(x)$ and $\mathcal{B}(x)$ are two GF of the two sequences $\langle a_n \rangle$ and $\langle b_n \rangle$, then their product $\mathcal{A}(x)$. $\mathcal{B}(x)$ is also a GF.*

Now, here, the generating function of Bernoulli plynomial of first kind, second kind, (for) is given by $G(x,t) = \frac{te^{xt}}{e^t - 1} = \sum_{n=0}^{\infty} \frac{B_n(x)}{n!} t^n$ and $G(x,t) = \frac{(1+t)^x}{\log(1+t)} t = \sum_{n=0}^{\infty} \frac{b_n(x)}{n!} t^n$, these comes under the category of exponential generating functions.

The GF of modified poly-Bernoulli numbers $C_n^{(k)}$ (see [32]), poly-Daehee numbers $D_n^{(k)}(x)$ (see [15]) are given as:

$$\frac{\mathrm{Li_k}(1-e^{-t})}{e^t-1}=\sum_{n=0}^{\infty}\frac{C_n^{(k)}}{n!}t^n, \qquad (|t|<\pi, \mathrm{k}\in\mathbb{Z}^+) \tag{6}$$

$$\frac{\log(1+t)}{\mathrm{Li_k}(1-e^{-t})}=\sum_{n=0}^{\infty}\frac{D_n^{(k)}}{n!}t^n, \qquad (|t|<\pi, \mathrm{k}\in\mathbb{Z}^+). \tag{7}$$

As we know that the Stirling numbers are the coefficients of the expansion of the falling and rising factorial, denoted by $(x)_n, (x)^n$, respectively. Falling and rising factorials are polynomials in degree n. The Stirling numbers of first kind $S_1(n,k)$ and the Stirling number of second kind $S_2(n,k)$ are generated by the following series expansion

$$\frac{(\log(1+t))^k}{k!}=\sum_{n=0}^{\infty}S_1(n,k)\frac{t^n}{n!} \text{ and } \frac{(e^t-1)^k}{k!}=\sum_{n=0}^{\infty}S_2(n,k)\frac{t^n}{n!}. \tag{8}$$

In view of Eq. (6), we can also obtain the following another connections of Stirling numbers.

$$x^{(n)}=\sum_{k=0}^{n}S_2(n,k)\,(x)_k, \tag{9}$$

where $S_2(n,0)=\delta_{n,0}, S_2(n,k)=0,$ for $k>n$, and $\delta_{n,k}$ is the kronecker delta. The Stirling number of the first kind, denoted by $S_1(n,k)$ is given as:

$$(x)_n=\sum_{l=0}^{n}S_1(n,l)x^l, \qquad (n\geq 0) \tag{10}$$

$(x)_n$ is called falling factorial, given by Eq. (3). After expanding falling factorial and right hand side of the above equation, we get list of Stirling numbers for different values of n.

Remark 2.1. *One of these falling and raising factorials can also be found by using*

$$(x)^n=(-1)^n(-x)_n$$

For $\mathrm{k}\in\mathbb{Z}$, $(\mathrm{k}>1)$, the classical polylogarithm function $\mathrm{Li_k}(x)$ (see [2, 30, 34]) is defined by

$$\mathrm{Li_k}(x)=\sum_{n=1}^{\infty}\frac{x^n}{n^{\mathrm{k}}}, \qquad (|x|<1). \tag{11}$$

Note 2.2. *For* $k=1, \mathrm{Li}_1(x)=\sum_{n=1}^{\infty}\frac{x^n}{n}=\frac{1}{\log(1-x)}$, *(here the sequuence* $\langle a_n\rangle=\frac{(n-1)!}{n!}$*)..*

In this chapter, we consider the poly-Changhee numbers and polynomials and derive new explicit formulas and identities for those numbers and polynomials.

3. Higher-order Changhee polynomials and numbers

As one of the special polynomial, the Changhee polynomials and numbers are closely related to some other important polynomials and numbers such as combinatorics, applied science, mathematical physics, etc. The generating function of the higher-order Changhee polynomials is described as:

$$\left(\frac{2}{t+2}\right)^k (1+t)^x = \sum_{n=0}^{\infty} Ch_n^{(k)}(x)\frac{t^n}{n!}. \tag{12}$$

where the sequence of the polynomial $Ch_n^{(k)}(x)$ is called a higher-order Changhee polynomial. For the different values of k, $(k \in \mathbb{N})$ and at $x = 0$, the constant terms of the Changhee polynomials are listed here:

$$Ch_n^{(0)}(0) = 1$$
$$Ch_n^{(1)}(0) = \frac{k}{2}$$
$$Ch_n^{(2)}(0) = \frac{k^2-k}{4}$$
$$Ch_n^{(3)}(0) = \frac{k^3-3k^2+4k}{8}$$
$$Ch_n^{(3)}(0) = \frac{k^4-4k^3+19k^2-28k}{16}$$
$$\cdots$$
$$\cdots$$

Note 3.1. *For* $k = 1$ *in* Eq. (10), *we get simple Changhee polynomials, given by the* Eq. (1).

Similarly, we can find other constant terms of the above set of Changhee Polynomial given by (10). Now, in general, some of the higher-order Changhee polynomials are also listed here (the generating function is given in (10)):

$$Ch_n^{(\mathrm{k})}(x) = 1$$
$$Ch_n^{(\mathrm{k})}(x) = x + \frac{k}{2}$$
$$Ch_n^{(\mathrm{k})}(x) = x^2 + (k-1)x + \frac{k^2-k}{4}$$
$$Ch_n^{(\mathrm{k})}(x) = x^3 + \left(\frac{k}{2}-1\right)x^2 + \left(\frac{8k^2-9k+8}{4}\right)x + \left(\frac{k^3-3k^2+4k}{8}\right)$$
$$Ch_n^{(\mathrm{k})}(x) = x^4 + (2k-6)x^3 + \left(\frac{3k^2-12k+22}{2}\right)x^2 + \left(\frac{k^3-3k^2+4k}{2}\right)x + \left(\frac{k^4-6k^3+19k^2-28k}{6}\right)$$
$$\cdots$$
$$\cdots$$

Note 3.2. *For different choices of k and n, we can easily find various simple Changhee polynomials and higher-order Changhee polynomials and numbers.*

In the next section, we shall define a simple poly-Changhee polynomial using the GF of the Changhee polynomial and the GF of the poly-logarithmic function.

4. Poly-Changhee polynomials and numbers

For $k \in \mathbb{Z}$, the poly-Changhee polynomials are defined by the following generating function to be

$$\left(\frac{2t}{2+t}\right)\frac{(1+t)^x}{\mathrm{Li_k}(1-e^{-t})} = \sum_{n=0}^{\infty} Ch_n^{(k)}(x)\frac{t^n}{n!}, \qquad (n \in \mathbb{N}_0). \tag{13}$$

For $x = 0, Ch_n^{(k)} = Ch_n^{(k)}(0)$ are called poly-Changhee numbers, given by

$$\frac{2t}{(2+t)\ \mathrm{Li_k}(1-e^{-t})} = \sum_{n=0}^{\infty} Ch_n^{(k)}(0)\frac{t^n}{n!}, \tag{14}$$

and by (12) we can easily obtain $Ch_0^{(k)} = 0$.
For the case $k = 1$, we have

$$\left(\frac{2t}{2+t}\right)\frac{(1+t)^x}{\mathrm{Li_1}(1-e^{-t})} = \frac{2(1+t)^x}{2+t} = \sum_{n=0}^{\infty} Ch_n^{(1)}(x)\frac{t^n}{n!}. \tag{15}$$

From (12) and (13), we get

$$Ch_n^{(1)}(x) = Ch_n(x), \qquad (n \geq 0). \tag{16}$$

Theorem 4.1. *Let $n \in \mathbb{Z}^* = \{0\} \cup \mathbb{Z}^+$, where $\mathbb{Z}^+$ denotes the positive integer. Then we have*

$$\begin{aligned} Ch_n^{(k)}(x) &= \sum_{l=0}^{n} Ch_l^{(k)}(x)_{n-l} \\ &= \sum_{l=0}^{n}(n-l)!\binom{n}{l}\binom{x}{n-l}Ch_l^{(k)}. \end{aligned} \tag{17}$$

Proof. Now from (11) the generating function of poly-Changhee polynomials, we have

$$\begin{aligned} \sum_{n=0}^{\infty} Ch_n^{(k)}(x)\frac{t^n}{n!} &= \left(\frac{2t}{2+t}\right)\frac{(1+t)^x}{\mathrm{Li_k}(1-e^{-t})} \\ &= \left(\frac{2t}{(2+t)\mathrm{Li_k}(1-e^{-t})}\right)(1+t)^x \\ &= \left(\sum_{l=0}^{\infty}\frac{Ch_l^{(k)}(0)t^l}{l!}\right)\left(\sum_{n=0}^{\infty}\frac{(x)_n t^n}{n!}\right) \\ &= \sum_{n=0}^{\infty}\sum_{l=0}^{\infty}\frac{Ch_l^{(k)}(0)(x)_n t^{n+l}}{n!l!} \\ &= \sum_{n=0}^{\infty}\sum_{l=0}^{n}\frac{Ch_l^{(k)}(0)(x)_{n-l}t^n}{(n-l)!l!} \\ &= \sum_{n=0}^{\infty}\left(\sum_{l=0}^{n}(n-l)!\binom{n}{l}Ch_l^{(k)}\binom{x}{n-l}\right)\frac{t^n}{n!}. \end{aligned} \tag{18}$$

Thus, we obtain the above result on equating the coefficients of t^n on both sides of Eq. (16).

Theorem 4.2. *Let* $n \in \mathbb{Z}^* = \{0\} \cup \mathbb{Z}^+$, *where* $\mathbb{Z}^+$ *denotes the positive integer. Then we have*

$$Ch_n^{(k)}(x) = \frac{Ch_{n+1}^{(k)}(x+1) - Ch_{n+1}^{(k)}(x)}{n+1}. \tag{19}$$

Proof. Again from (11) the generating function of poly-Changhee polynomials, we have

$$\begin{aligned}
&\sum_{n=0}^{\infty}\left(Ch_n^{(k)}(x+1) - Ch_n^{(k)}(x)\right)\frac{t^n}{n!} \\
&= \left(\frac{2t}{2+t}\right)\frac{(1+t)^{x+1}}{\mathrm{Li}_k(1-e^{-t})} - \left(\frac{2t}{2+t}\right)\frac{(1+t)^{x}}{\mathrm{Li}_k(1-e^{-t})} \\
&= t\left(\frac{2t}{2+t}\right)\frac{(1+t)^{x}}{\mathrm{Li}_k(1-e^{-t})} \\
&= \sum_{n=0}^{\infty} Ch_n^{(k)}(x)\frac{t^{n+1}}{n!}.
\end{aligned} \tag{20}$$

Thus, we acquire the proof of the theorem on equating the coefficients of t^{n+1} on both sides of Eq. (18).

Theorem 4.3. *Let* $n \in \mathbb{Z}^* = \{0\} \cup \mathbb{Z}^+$, *where* $\mathbb{Z}^+$ *denotes the positive integer. Then we have*

$$Ch_n^{(k)}(x+y) = \sum_{l=0}^{n}\binom{n}{l} Ch_{n-l}^{(k)}(x)\,(y)_l. \tag{21}$$

Proof. From (11) the generating function of poly-Changhee polynomials, we have

$$\begin{aligned}
\sum_{n=0}^{\infty} Ch_n^{(k)}(x+y)\frac{t^n}{n!} &= \left(\frac{2t}{2+t}\right)\frac{(1+t)^{x+y}}{\mathrm{Li}_k(1-e^{-t})} \\
&= \left(\left(\frac{2t}{2+t}\right)\frac{(1+t)^{x}}{\mathrm{Li}_k(1-e^{-t})}\right)(1+t)^y \\
&= \left(\sum_{n=0}^{\infty} Ch_n^{(k)}(x)\frac{t^n}{n!}\right)\left(\sum_{l=0}^{\infty}\frac{(y)_l t^l}{l!}\right) \\
&= \sum_{n=0}^{\infty}\sum_{l=0}^{\infty}\frac{Ch_n^{(k)}(x)(y)_l t^{n+l}}{l!\ n!} \\
&= \sum_{n=0}^{\infty}\sum_{l=0}^{n}\frac{Ch_{n-l}^{(k)}(x)(y)_l t^{n}}{l!\ (n-l)!} \\
&= \sum_{n=0}^{\infty}\sum_{l=0}^{n}\left(\binom{n}{l} Ch_{n-l}^{(k)}(x)\ (y)_l\right)\frac{t^n}{n!}
\end{aligned} \tag{22}$$

Thus, we acquire the above theorem on equating the coefficient of t^n on both sides of Eq. (20).

Theorem 4.4. For $n \geq 0$,

$$\sum_{l=0}^{n}\binom{n}{l} Ch_{n-l}^{(\mathrm{k})}(x)\ \mathrm{C}_l^{(k)} = \sum_{l=0}^{n}\binom{n}{l} Ch_{n-l}^{k}(x)\mathcal{B}_l$$

For every integer k poly-Bernoulli number $\mathfrak{B}_{\mathrm{n}}^{(\mathfrak{k})}$ and the modified poly-Bernoulli numbers $\mathrm{C}_{\mathrm{n}}^{(\mathrm{k})}$ introduced by Kaneko [23].

$$\frac{\mathrm{Li}_{\mathrm{k}}(1-e^{-t})}{1-e^{-t}} = \sum_{n=0}^{\infty}\frac{\mathfrak{B}_{\mathrm{n}}^{(\mathfrak{k})}}{n!}t^n \tag{23}$$

$$\frac{\mathrm{Li}_{\mathrm{k}}(1-e^{-t})}{e^{t}-1} = \sum_{n=0}^{\infty}\frac{\mathrm{C}_{\mathrm{n}}^{(\mathrm{k})}}{n!}t^n \tag{24}$$

Now, from the definition of poly-changhee polynomials, we observe that

$$\left(\frac{2t}{2+t}\right)\frac{(1+t)^x}{e^t-1} = \left(\frac{(2t)(1+t)^x}{(2+t)(\mathrm{Li}_{\mathrm{k}}(1-e^{-t}))}\right)\frac{\mathrm{Li}_{\mathrm{k}}(1-e^{-t})}{e^t-1} \tag{25}$$

$$\begin{aligned}
&= \left(\sum_{n=0}^{\infty}\frac{Ch_n^{(\mathrm{k})}(x)}{n!}t^n\right)\left(\sum_{l=0}^{\infty}\frac{\mathrm{C}_l^{(k)}}{l!}t^l\right)\\
&= \sum_{n=0}^{\infty}\sum_{l=0}^{\infty}\frac{Ch_n^{(\mathrm{k})}(x)\mathrm{C}_l^{(k)_l}t^{n+l}}{l!\ n!}\\
&= \sum_{n=0}^{\infty}\sum_{l=0}^{n}\frac{Ch_{n-l}^{(\mathrm{k})}(x)\mathrm{C}_l^{(k)}}{l!\ (n-l)!}\frac{t^n}{n!}(n!)\\
&= \sum_{n=0}^{\infty}\left(\sum_{l=0}^{n}\binom{n}{l} Ch_{n-l}^{(\mathrm{k})}(x)\ \mathrm{C}_l^{(k)}\right)\frac{t^n}{n!}
\end{aligned} \tag{26}$$

Now, left-hand side of the Eq. (24), we have

$$\left(\frac{2t}{2+t}\right)\frac{(1+t)^x}{e^t-1} = \left(\frac{2(1+t)^x}{2+t}\right)\left(\frac{t}{e^t-1}\right) \tag{27}$$

$$\begin{aligned}
&= \left(\sum_{n=0}^{\infty}Ch_n^{(\mathrm{k})}(x)\frac{t^n}{n!}\right)\left(\sum_{l=0}^{\infty}\mathcal{B}_l\frac{t^l}{l!}\right)\\
&= \sum_{n=0}^{\infty}\sum_{l=0}^{n}\frac{Ch_n^{(\mathrm{k})}(x)\ \mathcal{B}_l}{(n-l)!\ l!}t^{n+l}\\
&= \sum_{n=0}^{\infty}\left(\sum_{l=0}^{n} {n \atop l}\ Ch_{n-l}^{(\mathrm{k})}(x)\mathcal{B}_l\right)\frac{t^n}{n!}
\end{aligned} \tag{28}$$

S. No.	Name of Polynomial	$\mathcal{A}(y)$	Generating Function
I.	Changhee polynomial [13]	$\left(\frac{2}{2+t}\right)(1+t)^x$	$\sum_{n=0}^{\infty}\frac{t^n}{n!}Ch_n(x) = \frac{2}{2+t}(1+t)^x$
II.	Higher-order Changhee polynomial [3]	$\left(\frac{2}{2+t}\right)^k(1+t)^x$	$\sum_{n=0}^{\infty}\frac{t^n}{n!}Ch_n^{(k)}(x) = \left(\frac{2}{2+t}\right)^k(1+t)^x$
III.	Degenerate Changhee polynomial [38]	$\left(\frac{2\lambda}{2\lambda+\log(1+\lambda t)}\right)\left(1+\log(1+\lambda t)^{\frac{1}{\lambda}}\right)^x$	$\sum_{n=0}^{\infty}\frac{t^n}{n!}Ch_n^{*}(x) = \left(\frac{2\lambda}{2\lambda+\log(1+\lambda t)}\right)(1+\log(1+\lambda t))^{\frac{x}{\lambda}}$
IV.	Higher-order degenerate Changhee polynomial [35]	$\left(\frac{2\lambda}{2\lambda+\log(1+\lambda t)}\right)^{(k)}\left(1+\log(1+\lambda t)^{\frac{1}{\lambda}}\right)^x$	$\sum_{n=0}^{\infty}\frac{t^n}{n!}Ch_n^{*,(k)}(x) = \left(\frac{2\lambda}{2\lambda+\log(1+\lambda t)}\right)(1+\log(1+\lambda t))^{\frac{x}{\lambda}}$
V.	Poly-Changhee polynomial	$\left(\frac{2t}{2+t}\right)\frac{(1+t)^x}{\mathrm{Li}_k(1-e^{-t})}$	$\left(\frac{2t}{2+t}\right)\frac{(1+t)^x}{\mathrm{Li}_k(1-e^{-t})} = \sum_{n=0}^{\infty}Ch_n^{(k)}(x)\frac{t^n}{n!}$

Table 1.
Similar members to the Changhee family $Ch_n(x)$.

From Eqs. (25) and (27), we have

$$\sum_{l=0}^{n}\binom{n}{l}Ch_{n-l}^{(k)}(x)\,C_l^{(k)} = \sum_{l=0}^{n}\binom{n}{l}Ch_{n-l}^{k}(x)\mathcal{B}_l \tag{29}$$

Therefore, we arrive the above result (**Table 1**).

5. Conclusion

In this chapter, we have studied Changhee polynomial and their various generalization. In the same way, we can also define degenerate poly-Changhee polynomial by the following generating function. For $k \in \mathbb{Z}$, the degenerate poly-Changhee polynomials are defined by the following generating function:

$$\left(\frac{2\lambda t}{2\lambda+\log(1+\lambda t)}\right)\frac{\left(1+\log(1+\lambda t)^{\frac{1}{\lambda}}\right)^x}{\mathrm{Li}_k(1-e^{-t})} = \sum_{n=0}^{\infty}Ch_{n,\lambda}^{(k)}(x)\frac{t^n}{n!}, \qquad (n \in \mathbb{N}_0). \tag{30}$$

Remark 5.1. The limiting case (i.e., when $\lim_{\lambda\to 0}$) of degenerate poly-Changhee polynomials gives poly-Changhee polynomials (for more details, visit [12, 31, 33, 36]).

$$\lim_{\lambda\to 0}\left\{\sum_{n=0}^{\infty} Ch_{n,\lambda}^{(\mathrm{k})}(x)\frac{t^n}{n!}\right\} = \lim_{\lambda\to 0}\left\{\left(\frac{2\lambda t}{2\lambda+\log(1+\lambda t)}\right)\frac{(1+\log(1+\lambda t))^{\frac{x}{\lambda}}}{\mathrm{Li_k}(1-e^{-t})}\right\},$$
$$= \lim_{\lambda\to 0}\left(\frac{2\lambda t}{2\lambda+\log(1+\lambda t)}\right)\lim_{\lambda\to 0}\left(\frac{(1+\log(1+\lambda t))^{\frac{x}{\lambda}}}{\mathrm{Li_k}(1-e^{-t})}\right)$$
$$= \left(\frac{2t}{2+t}\right)\frac{(1+t)^x}{\mathrm{Li_k}(1-e^{-t})}$$

In the same way, we can also define higher-order degenerate poly-Changhee polynomial by the following generating function:

$$\left(\frac{2\lambda t}{2\lambda+\log(1+\lambda t)}\right)^{\alpha}\frac{\left(1+\log\left(1+\lambda t\right)^{\frac{1}{\lambda}}\right)^{x}}{\mathrm{Li_k}(1-e^{-t})} = \sum_{n=0}^{\infty} Ch_{n,\alpha,\lambda}^{(\mathrm{k})}(x)\frac{t^n}{n!}, \tag{31}$$

where, $n \in \mathbb{N}_0, \mathrm{k} \in \mathbb{Z}$. After substituting $x = 0$ in Eq. (30), we get higher-order degenerate poly-Changhee numbers, given as:

$$\left(\frac{2\lambda t}{2\lambda+\log(1+t)}\right)^{\alpha}\frac{1}{\mathrm{Li_k}(1-e^{-t})} = \sum_{n=0}^{\infty} Ch_{n,\alpha,\lambda}^{(\mathrm{k})}(0)\frac{t^n}{n!}. \tag{32}$$

Thus, one can see easily the more general results have been obtianed in this chapter.

Open Question: The authors suggest that the readers try to think about how to find generating function of various types of Apostol-based poly-Changhee polynomials, Bernoulli-based poly-Changhee polynomials, and Euler-based poly-Changhee polynomials?

Classifications

2010 Mathematics Subject Classification, 33C20, 33C45, 65A05.

Author details

Mohd. Saif[1] and Raghib Nadeem[2*]

1 Department of Mathematics, Madanpalli Institute of Technology and Science, Madanpalli, India

2 Department of Applied Sciences, Maulana Mukhtar Ahmad Nadvi Technical Campus, Malegaon, India

*Address all correspondence to: raghibmaths2015@gmail.com

References

[1] Kim T, Kim DS, Mansour T, Rim S-H, Schork M. Umbral calculus and Sheffer sequences of polynomials. Journal of Mathematical Physics. 2013;**54**(8):083504

[2] Kim DS, Kim T, Lee SH. A note on poly-Bernoulli polynomials arising from umbral calculus. Advanced Studies in Theoretical Physics. 2013;**7**(15):731-744

[3] Kim D, S. Higher-order Changhee numbers and polynomials. Advanced Studies in Theoretical Physics. 2014; **8**(8):365-373

[4] Rim S-H, Park J-W, Pyo S-S, Kwon J. The n-th twisted Changhee polynomials and numbers. Bulletin of the Korean Mathematical Society. 2015;**52**(3): 741-749

[5] Jang, Lee-Chae and Ryoo, Cheon Seoung and Seo, Jong-Jin, Kwon, Hyuck In, Some properties of the twisted Changhee polynomials and their zeros. Applied Mathematics and Computation. 2016;**274**:169–177

[6] Kim T, Dolgy DV, Kim DS, Seo JJ. Differential equations for Changhee polynomials and their applications. Journal of Nonlinear Sciences and Applications. 2016;**9**:2857-2864

[7] Kim T, Kim DS. A note on nonlinear Changhee differential equations. Russian Journal of Mathematical Physics. 2016; **23**(1):88-92

[8] Arakawa T, Kaneko M, et al. On poly-Bernoulli numbers. Commentarii Mathematici Universitatis Sancti Pauli. 1999;**48**(2):159-168

[9] Bayad A, Hamahata Y. Polylogarithms and poly-Bernoulli polynomials. Kyushu Journal of Mathematics. 2011;**65**(1):15-24

[10] Bayad Abdelmejid, Hamahata Yoshinori, et al. Multiple polylogarithms and multi-poly-Bernoulli polynomials. Functiones et Approximatio Commentarii Mathematici. 2012;**46**(1):45–61

[11] Bhargava M. The factorial function and generalizations. The American Mathematical Monthly. 2000;**107**(9): 783-799

[12] Kim BM, Kim T, Park J-W, Radwan TA. Identities on Changhee polynomials arising from λ-Sheffer sequences. Complexity. 2022;**2022**: 5868689. 16 pages. DOI: 10.1155/2022/5868689

[13] Kim DS. A note on Changhee polynomials and numbers. Advanced Studies in Theoretical Physics. 2013; 7(20):993-1003

[14] Ding D, Yang J. Some identities related to the Apostol-Euler and Apostol-Bernoulli polynomials. Advanced Studies in Contemporary Mathematics. 2010; **20**(1):7-21

[15] Lim D, Kwon J. Proceedings of the Jangjeon Mathematical Society. 2016; **19**(2):219-224

[16] Dattoli G, Lorenzutta S, Cesarano C. Finite sums and generalized forms of Bernoulli polynomials. Rend. Math. 1999;**19**:385-391

[17] Horadam AF. Falling factorial polynomials of generalized Fibonacci type. In: Applications of Fibonacci Numbers: Volume 3 Proceedings of 'The Third International Conference on Fibonacci Numbers and Their Applications', Pisa, Italy, July 25-29, 1988. Netherlands: Springer; 1990. pp. 139-153

[18] Hamahata Y et al. Poly-Euler polynomials and Arakawa–Kaneko type zeta functions. Functiones et Approximatio Commentarii Mathematici. 2014;**51**(1):7-22

[19] Jolany H, Darafsheh MR, Alikelaye RE. Generalizations of poly-Bernoulli numbers and polynomials. International Journal of Mathematical Combinatorics. 2010;**2**:7-14

[20] Jolany H, Aliabadi M, Corcino RB, Darafsheh MR. A note on multi poly-Euler numbers and Bernoulli polynomials. arXiv preprint arXiv: 1401.2645. 2014 Jan 12

[21] Gould HW. Explicit formulas for Bernoulli numbers. The American Mathematical Monthly. 1972;**79**:44-51

[22] Kaneko M. Poly-bernoulli numbers. Journal de Théorie des Nombres de Bordeaux. 1997;**9**(1):221-228

[23] Kaneko, Masanobu par, Poly-Bernoulli numbers, Journal de Théori des Nomres de Bordeaux. 1997;**9**(1):221-228

[24] Kim, Taekyun and Jang, Yu Seon and Seo, Jong Jin a note on poly-Genocchi numbers and polynomials, Applied Mathematical Sciences. 2014; **8**(96):4775-4781

[25] Kim DS, Kim T, Seo JJ. A note on Changhee polynomials and numbers. Advanced Studies in Theoretical Physics. 2013;**7**(20):993-1003

[26] Kim, Taekyun and San Kim, Dae, Identities of symmetry for degenerate Euler polynomials and alternating generalized falling factorial sums, Iranian Journal of Science and Technology, Transactions A: Science. 2017;**41**(4):939-949

[27] Kim D, Kim T. A note on poly-Bernoulli and higher-order poly-Bernoulli polynomials. Russian Journal of Mathematical Physics. 2015;**22**(1):26-33

[28] Kim DS, Kim T. Poly-Cauchy numbers and polynomials with umbral calculus viewpoint. arXiv preprint arXiv: 1307.5188. 2013 Jul 19

[29] Kim Taekyun, Kwon Hyuck In, Lee Sang Hun, Seo, Jong Jin. A note on poly-Bernoulli numbers and polynomials of the second kind.Advances in Difference Equations. 2014;**2014**(1):219

[30] Kim DS, Kim T, Rim SH. Umbral calculus and Euler polynomials. arXiv preprint arXiv: 1211.6639. 2012 Nov 22

[31] Kim T, Kim D, Kim HY, et al. Degenerate poly-Bernoulli polynomials arising from degenerate polylogarithm. Advances in Difference Equations. 2020; **444**:444. DOI: 10.1186/s13662-020-02901-9

[32] Kaneko M. Poly Bernoulli numbers. Journal de Théorie des Nombres de Bordeaux. 1997;**9**:199-206

[33] Saif M, Nadeem R. Evaluation of Apostol Euler based poly Daehee polynomials. International Journal of Applied and Computational Mathematics. 2020;**6**:1. DOI: 10.1007/s40819-019-0748-2

[34] Dae SK, Taekyun K. Higher-order Frobenius-Euler and poly-Bernoulli mixed-type polynomials. Advances in Difference Equations. 2013;**2013**(1):251

[35] Yun SJ, Park J-W. Identities of degenerate poly-Changhee polynomials Arisingfrom λ-Sheffer sequences. Hindawi Journal of Mathematics. 2022;**9**: 1482534. DOI: 10.1155/2022/1482534

[36] Jin S, Dagli MC, Qi F. Degenerate Fubini-type polynomials and numbers, degenerate Apostol–Bernoulli

polynomials and numbers, and degenerate Apostol–Euler polynomials and numbers. Axioms. 2022;**11**:477

[37] Son J-W, Kim M-S. On poly-Eulerian numbers. Bulletin of the Korean Mathematical Society. 1999;**36**(1):47-61

[38] Kim T, Kim DS. Degenerate Changhee numbers and polynomials of the second kind. arXiv preprint arXiv: 1707.09721. 2017 Jul 31

[39] Kim T, Kim DS, Kim HK. On generalized degenerate Euler–Genocchi polynomials. Applied Mathematics in Science and Enginnering. 2022;**31**(1): 2159958

[40] Ohno Y, Sasaki Y. On poly-Euler numbers. Journal of the Australian Mathematical Society. 2017;**103**(1): 126-144

[41] Wang YX, Smola A, Tibshirani R. The falling factorial basis and its statistical applications. In: International Conference on Machine Learning. PMLR. Jun 2014. pp. 730-738

Chapter 10

Existence of Limit Cycles for a Class of Quintic Kukles Homogeneous System

Sarah Abdullah Qadha, Muneera Abdullah Qadha and Haibo Chen

Abstract

We investigated the existence of limit cycles for quintic Kukles polynomial differential systems depending on a parameter in this chapter. These systems are important in practical applications and theoretical advances. We first used formal series method based on Poincaré's ideas to prove this point and determine the center-focus problem. We then utilized the Dulac function to prove the nonexistence of closed orbits. We determined the sufficient condition for the existence of the limit cycles, which bifurcate from the equilibrium point, using Hopf bifurcation theory. Lastly, we provided some numerical examples for illustration using MATLAB to plot. Note that studies on the existence and the nonexistence of limit cycles and algebraic limit cycles for Kukles systems are limited.

Keywords: singular point, center-focus, existence, limit cycle, Kukles

1. Introduction

Limit cycle discovery by Poincaré in his paper Integral Curves defined by differential equations (1881–1886) [1–4]. Physicists [5] failed to describe the oscillation phenomenon through the linear differential equation in the twentieth century. Van der Pol [6] proposed van der Pol equation in 1926 to describe the oscillation of the constant amplitude of a triode vacuum tube.

The limit cycle has attracted the attention of many pure and applied mathematicians. Numerous mathematical models from physics, biology, economics, engineering, and chemistry have been proposed since the 1950s to explore autonomous plane systems with a limit cycle [5, 7]. Limit cycle theory is closely related to Hilbert's 16th problem. Exploring the existence of limit cycles is crucial in this theory. Poincaré proposed the method of topographical system, the successor function, small parameter method, and the annular region theorem to determine the existence of limit cycles. Existence, nonexistence, uniqueness, and other properties of the limit cycle have been extensively analyzed by mathematicians and physicists [8].

The problem of the existence of periodic solutions in the Liénard equation was explored in a previous study [9].

$$\ddot{x} + f(x)\dot{x} + g(x) = 0. \tag{1}$$

This problem has been widely investigated since the study of Liénard. The Liénard equation in the phase plane is equivalent to the system

$$\begin{aligned} \frac{dx}{dt} &= y, \\ \frac{dy}{dt} &= -g(x) - f(x)y. \end{aligned} \tag{2}$$

The Liénard equation in the Liénard plane is equivalent to the system and can be expressed as follows:

$$\begin{aligned} \frac{dx}{dt} &= y - F(x), \\ \frac{dy}{dt} &= -g(x), \end{aligned} \tag{3}$$

where $F(x) = \int_0^x f(x)dx$. On this basis, the problem of the existence of periodic solutions is brought back to a problem of the existence of limit cycles for the previous systems (3).

For the more general equation

$$\ddot{x} + f(x, \dot{x})\dot{x} + g(x) = 0. \tag{4}$$

We observed that the Liénard system (3) has become invalid because f depends on two variables. Meanwhile, the phase plane system (2) can be transformed into the following formula:

$$\begin{aligned} \frac{dx}{dt} &= y, \\ \frac{dy}{dt} &= -g(x) - f(x, y)y. \end{aligned} \tag{5}$$

Accordingly, several limit cycles exist [9]. The problem of limit cycles of the Eq. (4) was first explored by Norman Levinson, Oliver K. Smith, and Dragilev [5].

Our work is related to the differential system of the following form:

$$\begin{aligned} \frac{dx}{dt} &= y, \\ \frac{dy}{dt} &= -x + P_n(x, y), \end{aligned} \tag{6}$$

where $P_n(x, y)$ is a homogeneous polynomial of the degree n. System (6) contains a center point at the origin if and only if this system is symmetric to one of the coordinate axes because $n \geq 2$. The Russian mathematician I. S. Kukles first investigated differential polynomial systems (6) in 1944. System (6) was then called the Kukles homogeneous systems [10–12]. The polynomial differential system (6) is linear when $n = 1$. Obtaining isolated periodic solutions in the set of all periodic solution linear differential systems is impossible. The polynomial differential system (6) is a quadratic system when $n = 2$, in which the system is symmetric to the y-axis; and this system has been extensively investigated [13]. The cubic system (6) was examined when n = 3 to obtain a system (6) that has six small amplitude limit cycles in the neighborhood of the origin [6, 14, 15]. Rebiha Benterki and Jaume Llibre examined system (6), which is symmetric to the $y - axis$, when n = 4 in 2017 and provides a sufficient condition for the existence of limit cycles [10]. The system (6) always contains either a center or a fine focus (weak focus) at the origin.

Consider the polynomial differential system

$$\frac{dx}{dt} = -y,$$
$$\frac{dy}{dt} = x + b_1x^2 + b_2xy + b_3y^2 + b_4x^3 + b_5x^2y + b_6xy^2 + b_7y^3. \quad (7)$$

Kukles provided the necessary and sufficient conditions for the origin to be centered. However, some new cases have been discovered. C. J. Christopher and N. G. Lloyd [16] explored the case $b_7 = 0$, (reduced Kukles system) in 1990 and established the system (7), which contains a maximum of five limit cycles bifurcating from the origin. N. G. Lloyd [17] and J. M. Pearson [18] analyzed the case $b_2 = 0$ in 1990 and established system (7), which contains a maximum six limit cycles bifurcating from the origin. The central problem is to characterize between a center and a focus at the origin of the system. If all orbits in the neighborhood spiral toward or away from the origin, then the origin is the focus. If all orbits in the neighborhood, except the origin, are periodic, then the origin is the center. This problem of characterizing between a center and a focus at the origin has been solved only in linear and quadratic systems. However, a few particular cases in families of high degree still require further investigation. Some studies have discussed the existence and the nonexistence of small amplitude limit cycles and algebraic limit cycles for Kukles systems. A previous study proved that system (7) with nonlinearities of a degree higher than two contains a center at the origin if and only if its vector field is symmetric to one of the coordinate axes in 2015 [11].

We present some necessary definitions and theorems in the first section of this chapter. We used a formal series method based on Poincaré's ideas to determine the center focus in the second section. By the Hopf bifurcation theory, we obtained the sufficient condition for existence of limit cycles for a following quintic Kukles polynomial differential system depending on a parameter

$$\frac{dx}{dt} = y,$$
$$\frac{dy}{dt} = -x - 2x^3 + b_{21}x^2y + b_{12}xy^2 + b_{50}x^5 + b_{41}x^4y, \quad (8)$$

where b_{21}, b_{12}, b_{50} and b_{41} are parameters, and x and y are real variables that bifurcate from the equilibrium point (singular point). Hence, we establish sufficient conditions for the existence of the limit cycle according to the change analysis of the stability of the focus when the parameters change. Lastly, we presented some numerical examples for illustration. Notably, system (8) contains several limit cycles and the Liénard system is used to prove that the existence is already invalid.

2. Some preliminary results

In this section, we introduce some definitions and notations regarding the existence of the quintic differential system (8).

2.1 Conjecture

The origin, except for the linear center, is not the isochronous center of system (7) if the system has odd degree. Therefore, if the linear term contains a center, and the degree of the nonlinear term is odd, then the origin cannot be the center simultaneously.

2.2 Local results for Liénard systems

Blows and Lloyd proved the following results for system (2) in 1989, where $f(x) = a_0 + a_1x + a_2x^2 + \ldots + a_mx^m$ and $g(x) = x + b_2x^2 + b_3x^3 + \ldots + b_nx^n$, m and n are the natural numbers. Let $H(m,n)$ denote the maximum numbers of small-amplitude limit cycles that can be bifurcated from the origin for system (2), where m is the degree of f and n is the degree of g. If g is odd and the order of $f = m = 2i$ *or* $2i+1$, then $H(m,n) = i$ [19].

3. Singular point for system

The singular points for system (8) are $\boldsymbol{A}(\mathbf{0},\mathbf{0})$,
$\boldsymbol{B}\left(\sqrt[2]{\left(\frac{1+\sqrt{1+b_{50}}}{b_{50}}\right)},\mathbf{0}\right)$, $\boldsymbol{C}\left(-\sqrt[2]{\left(\frac{1+\sqrt{1+b_{50}}}{b_{50}}\right)},\mathbf{0}\right)$, $\boldsymbol{D}\left(\sqrt[2]{\left(\frac{1-\sqrt{1+b_{50}}}{b_{50}}\right)},\mathbf{0}\right)$ and $\boldsymbol{E}\left(-\sqrt[2]{\left(\frac{1-\sqrt{1+b_{50}}}{b_{50}}\right)},\mathbf{0}\right)$.

Remark 1

1. The following cases are presented for singular points B *and* C :
   If $b_{50} = -1$, then points B *and* $C \in \mathbb{C}$, and $\mathbb{C}$ is a complex number.
   If $b_{50} < -1$, then points B *and* $C \in \mathbb{C}$.
   If $-1 < b_{50} < 0$, then the points B *and* $C \in \mathbb{C}$.
   If $b_{50} > 0$, then points B *and* $C \in \mathbb{R}$ (real number).

2. The following cases are presented for singular points D *and* E :
   If $b_{50} = -1$, then points D *and* $E \in \mathbb{C}$.
   If $b_{50} > 1$, then points D *and* $E \in \mathbb{C}$.
   If $b_{50} < -1$, then points D *and* $E \in \mathbb{C}$.
   If $0 < b_{50} < 1$, then points D *and* $E \in \mathbb{C}$.
   If $-1 < b_{50} < 0$, then points D *and* $E \in \mathbb{C}$.

Note: Singular points D *and* E are not investigated in this study because we consider x *and* y as real variables.

Note: The linear system contains the center [20]. However, the perturbation of these centers inside the class of the linear differential system fails to produce a limit cycle because a linear differential system cannot contain an isolated periodic solution in the set of all periodic solutions.

4. Nonlinear system

For system (8) the associated nonlinear system gave by calculating the following Jacobian matrix:

$$J(x,y) = \begin{pmatrix} 0 & 1 \\ -1 - 6x^2 + 2b_{21}xy + b_{12}y^2 + 5b_{50}x^4 + 4b_{41}x^3y & b_{21}x^2 + 2b_{12}xy + b_{41}x^4 \end{pmatrix}$$

The characteristic equation is

$$= \lambda^2 - \left(b_{21}x^2 + 2b_{12}xy + b_{41}x^4\right)\lambda - \left(-1 - 6x^2 + 2b_{21}xy + b_{12}y^2 + 5b_{50}x^4 + 4b_{41}x^3y\right) = 0$$

Let $Tr = -\left(b_{21}x^2 + 2b_{12}xy + b_{41}x^4\right)$,

$$det = -\left(-1 - 6x^2 + 2b_{21}xy + b_{12}y^2 + 5b_{50}x^4 + 4b_{41}x^3y\right),$$

then the characteristic equation is $D(\lambda) = \lambda^2 + Tr\lambda + det = 0$.
Its roots are

$$\lambda_1, \lambda_2 = \frac{-Tr \mp \sqrt{Tr^2 - 4det}}{2}$$

1. For singular point $\boldsymbol{B}\left(\sqrt[2]{\left(\frac{1+\sqrt{1+b_{50}}}{b_{50}}\right)}, \mathbf{0}\right)$ of system (8) and $b_{50} > 0$, the Jacobian matrix

$$J\left(\sqrt[2]{\left(\frac{\mathbf{1} + \sqrt{\mathbf{1} + \boldsymbol{b_{50}}}}{\boldsymbol{b_{50}}}\right)}, \mathbf{0}\right) = \begin{pmatrix} 0 & 1 \\ -1-6x_0^2+5b_{50}x_0^4 & b_{21}x_0^2+b_{41}x_0^4 \end{pmatrix}$$

where $x_0 = \sqrt[2]{\left(\frac{1+\sqrt{1+b_{50}}}{b_{50}}\right)}$

The characteristic equation is

$$\lambda^2 - \left(b_{21}x_0^2 + b_{41}x_0^4\right)\lambda - \left(-1 - 6x_0^2 + 5b_{50}x_0^4\right) = 0$$

Its roots are

$$\lambda_{1,2} = \frac{\left(b_{21}x_0^2 + b_{41}x_0^4\right) \pm \sqrt{\left(b_{21}x_0^2 + b_{41}x_0^4\right)^2 + 4\left(-1 - 6x_0^2 + 5b_{50}x_0^4\right)}}{2}$$

$\left(-1 - 6x_0^2 + 5b_{50}x_0^4\right)$ can be written as $5b_{50}y^2 - 6y - 1 = 0$. Then

$$y_{1,2} = \frac{6 \pm \sqrt{36 + 20b_{50}}}{10b_{50}}$$

When $36 + 20b_{50} > 0$ and $b_{50} > 0$, then the two roots $\lambda_{1,2}$ are real roots with different signs. Then the singular point $\boldsymbol{B}\left(\sqrt[2]{\left(\frac{1+\sqrt{1+b_{50}}}{b_{50}}\right)}, \boldsymbol{0}\right)$ is a saddle.

2. For singular point $\boldsymbol{C}\left(-\sqrt[2]{\left(\frac{1+\sqrt{1+b_{50}}}{b_{50}}\right)}, \boldsymbol{0}\right)$ of system (8) when and $b_{50} \neq 0$, the Jacobian matrix

$$J\left(-\sqrt[2]{\left(\frac{1+\sqrt{1+b_{50}}}{b_{50}}\right)}, \boldsymbol{0}\right) = \begin{pmatrix} 0 & 1 \\ -1 - 6x_0^2 + 5b_{50}x_0^4 & b_{21}x_0^2 + b_{41}x_0^4 \end{pmatrix}$$

The characteristic equation is

$$\lambda^2 - (b_{21}x_0^2 + b_{41}x_0^4)\lambda - (-1 - 6x_0^2 + 5b_{50}x_0^4) = 0$$

where $x_0 = -\sqrt[2]{\left(\frac{1+\sqrt{1+b_{50}}}{b_{50}}\right)}$

Its roots are

$$\lambda_{1,2} = \frac{(b_{21}x_0^2 + b_{41}x_0^4) \pm \sqrt{(b_{21}x_0^2 + b_{41}x_0^4)^2 + 4(-1 - 6x_0^2 + 5b_{50}x_0^4)}}{2}$$

$(-1 - 6x_0^2 + 5b_{50}x_0^4)$ can be rewritten as $5b_{50}y^2 - 6y - 1 = 0$, then

$$y_{1,2} = \frac{6 \pm \sqrt{36 + 20b_{50}}}{10b_{50}}$$

When $36 + 20b_{50} > 0$ and $b_{50} > 0$, then the two roots $\lambda_{1,2}$ are real roots with different signs. Then the singular point $\boldsymbol{C}\left(-\sqrt[2]{\left(\frac{1+\sqrt{1+b_{50}}}{b_{50}}\right)}, \boldsymbol{0}\right)$, is a saddle.

3. The Jacobian matrix for the singular point $A(0,0)$ of system (8) is $J(0,0) = \begin{pmatrix} 0 & 1 \\ -1 & 0 \end{pmatrix}$

Its characteristic equation is $\lambda^2 + 1 = 0 \Rightarrow \lambda = \pm i$; then the singular point for the nonlinear system (8) is difficult to distinguish whether the singular point $A(0,0)$ is a center or a focus, so we will use the formal series method to decide it. We will show the detailed process in the next section.

5. Center-focus

Determining the center focus in the qualitative theory of differential equations, especially for the plane of the high-order polynomial differential system, is difficult and troublesome. According to the Hopf bifurcation theory, when we analyze the conditions of the limit cycle branching from the equilibrium point and the stability of the generated cycle, we must make a detailed analysis of the central focus. The qualitative analysis of a class of high-order differential systems is presented in this chapter. It is obvious that $A(0,0)$ is the center of the linear system corresponding to the system (8). We need to determine the center focus problem of singularity $A(0,0)$ of the nonlinear system. We use the formal series method based on Poincaré's ideas to examine the behavior of singularity $A(0,0)$.

6. Main results

6.1 Determination of the center-focus

Theorem 1 The following are assumed for system (8):

1. If $b_{21} > 0$, then the point $A(0,0)$ is the first-order unstable weak focus.
2. If $b_{21} < 0$, then the point $A(0,0)$ is the first-order stable weak focus.
3. If $b_{21} = 0, b_{41} > 0$, then the point $A(0,0)$ is the second-order unstable weak focus.
4. If $b_{21} = 0, b_{41} < 0$, then the point $A(0,0)$ is the second-order stable weak focus
5. If $b_{21} = 0, b_{41} = 0$, then the point $A(0,0)$ is the center.

Proof Suppose the system (8) has a solution of the following series form, then

$$F(x,y) = x^2 + y^2 + \sum_{k=3}^{\infty} F_k(x,y),$$

where F_k is the k-degree homogeneous polynomial of x and y. Then,

$$\begin{aligned}\left.\frac{dF}{dt}\right|_{(8)} &= \frac{\partial F}{\partial x}\frac{dx}{dt} + \frac{\partial F}{\partial y}\frac{dy}{dt} \\ &= \left(2x + \sum_{k=3}^{\infty}\frac{\partial F_x}{\partial x}\right)(y) \\ &\quad + \left(2y + \sum_{k=3}^{\infty}\frac{\partial F_x}{\partial y}\right)(-x - 2x^3 + b_{21}x^2y \\ &\quad + b_{12}xy^2 + b_{50}x^5 + b_{41}x^4y)\end{aligned} \tag{9}$$

At the right end of the above equation, starting from the third term, by making the homogeneous equations of the same order equal to zero, we can obtain series equations as follows:

Let the third power term at the right end of the Formula (9) be zero. We obtain

$$x\frac{\partial F_3}{\partial y} - y\frac{\partial F_3}{\partial x} = 2(xP_2 + yQ_2) = 0$$

We take the polar coordinate of the above formula $x = r\cos\theta, y = r\sin\theta, r^2 = x^2 + y^2$ and remove r^3.

$$\frac{dF_3(\cos\theta,\ \sin\theta)}{d\theta} = 0,$$

$$F_3(\cos\theta,\ \sin\theta) = constant,$$

$$F_3(x,y) = constant.$$

Let the term of power four at the right end of the formula (9) be zero, we find

$$\begin{aligned} x\frac{\partial F_4}{\partial y} - y\frac{\partial F_4}{\partial x} &= 2(xP_3 + yQ_3) + \left(P_2\frac{\partial F_3}{\partial x} + Q_2\frac{\partial F_3}{\partial y}\right) \\ &= -4x^3y + 2b_{21}x^2y^2 + 2b_{12}xy^3. \end{aligned}$$

Similarly, we use polar coordinates to find F_4 and remove r^4 to obtain

$$\begin{aligned} D_4 = \frac{dF_4(\cos\theta,\ \sin\theta)}{d\theta} &= -4\cos^3\theta\sin\theta + 2b_{21}\cos^2\theta\sin^2\theta \\ &\quad +2b_{12}\cos\theta\sin^3\theta. \end{aligned}$$

We discuss three situations as follows:

First case: $b_{21} \neq 0$.

Since

$$\int_0^{2\pi} -4\cos^3\theta\sin\theta + 2b_{21}\cos^2\theta\sin^2\theta + 2b_{12}\cos\theta\sin^3\theta d\theta \neq 0.$$

We take F_4 to satisfy the equation

$$\frac{dF_4(\cos\theta,\ \sin\theta)}{d\theta} = D_4 - C_4,$$

where

$$C_4 = \frac{b_{21}}{\pi}\int_0^{2\pi} \cos^2\theta\sin^2\theta d\theta.$$

C_4 has the same sign as b_{21} . Thus, $C_4 = \frac{b_{21}}{4}$. If

$$V(x,y) = x^2 + y^2 + F_3 + F_4.$$

Then,

$$\left.\frac{dV}{dt}\right|_{(8)} = C_4 r^4 + O(r^4).$$

When $b_{21} > 0$, *the point* $A(0,0)$ is the first-order unstable weak focus.
When $b_{21} < 0$, *the point* $A(0,0)$ is the first-order stable weak focus.
Second case: $b_{21} = 0, b_{41} \neq 0$. Thus,

$$\frac{dF_4(\cos\theta,\ \sin\theta)}{d\theta} = -4\,cos^3\theta\sin\theta + 2b_{12}\cos\theta\, sin^3\theta,$$

$$F_4(\cos\theta,\ \sin\theta) = cos^4\theta + \frac{b_{12}\,sin^4\theta}{2},$$

$$F_4(x,y) = x^4 + \frac{b_{12}}{2}y^4.$$

Simplify (note that $b_{21} = 0$)
Let the term of power five at the end of the formula (9) be zero, then

$$\begin{aligned}
x\frac{\partial F_5}{\partial y} - y\frac{\partial F_5}{\partial x} &= 2(xP_4 + yQ_4) + \sum_{k=3}^{4}\left(P_{5-k+1}\frac{\partial F_k}{\partial x} + Q_{5-k+1}\frac{\partial F_k}{\partial y}\right),\\
&= 2(xP_4 + yQ_4) + \left(P_3\frac{\partial F_3}{\partial x} + Q_3\frac{\partial F_3}{\partial y}\right) + \left(P_2\frac{\partial F_4}{\partial x} + Q_2\frac{\partial F_4}{\partial y}\right),\\
&= \left(-2x^3 + b_{21}x^2y + b_{12}xy^2\right)\frac{\partial F_3}{\partial y}.
\end{aligned}$$

Given that $F_3 = constant$, then $\frac{\partial F_3}{\partial y} = 0$. Thus,

$$x\frac{\partial F_5}{\partial y} - y\frac{\partial F_5}{\partial x} = 0.$$

We take the polar coordinate of the above and remove r^5 to obtain

$$\frac{dF_5(\cos\theta,\ \sin\theta)}{d\theta} = 0,$$

$$F_5(\cos\theta,\ \sin\theta) = constant,$$

$$F_5(x,y) = constant.$$

Simplify (note that $b_{21} = 0$)
Let the term of power six at the end of the formula (9) be zero, then

$$\begin{aligned}
x\frac{\partial F_6}{\partial y} - y\frac{\partial F_6}{\partial x} &= 2(xP_5 + yQ_5) + \sum_{k=3}^{5}\left(P_{6-k+1}\frac{\partial F_k}{\partial x} + Q_{6-k+1}\frac{\partial F_k}{\partial y}\right),\\
&= 2(xP_5 + yQ_5) + \left(P_4\frac{\partial F_3}{\partial x} + Q_4\frac{\partial F_3}{\partial y}\right),
\end{aligned}$$

$$+\left(P_3\frac{\partial F_4}{\partial x}+Q_3\frac{\partial F_4}{\partial y}\right)+\left(P_2\frac{\partial F_5}{\partial x}+Q_2\frac{\partial F_5}{\partial y}\right),$$
$$=2b_{50}x^5y+2b_{41}x^4y^2+\left(-2x^3+b_{12}x\,y^2\right)\frac{\partial F_4}{\partial y}.$$

When $b_{21}=0,\ F_4(x,y)=x^4+\frac{b_{12}}{2}y^4,\ \frac{\partial F_4}{\partial y}=2\,b_{12}y^3$ .Thus,

$$x\frac{\partial F_6}{\partial y}-y\frac{\partial F_6}{\partial x}=2b_{50}x^5y+2b_{41}x^4y^2-4b_{12}x^3y^3+2b_{12}^2xy^5.$$

We take the polar coordinate of the above equation and remove r^6 to obtain

$$D_6=\frac{dF_6(\cos\theta,\ \sin\theta)}{d\theta}=2b_{50}\cos^5\theta\sin\theta+2b_{41}\cos^4\theta\sin^2\theta$$
$$-4b_{12}\cos^3\theta\sin^3\theta+2b_{12}^2\cos\theta\sin^5\theta.$$

When $b_{41}\neq 0$,

$$\int_0^{2\pi}2b_{50}\cos^5\theta\sin\theta+2b_{41}\cos^4\theta\sin^2\theta-4b_{12}\cos^3\theta\sin^3\theta+2b_{12}^2\cos\theta\sin^5\theta d\theta\neq 0.$$

We take F_6 to satisfy the equation as follows:

$$\frac{dF_6(\cos\theta,\ \sin\theta)}{d\theta}=D_6-C_6,$$

where $C_6=\frac{b_{41}}{\pi}\int_0^{2\pi}\cos^4\theta\sin^2\theta d\theta,\ C_6=\frac{b_{41}}{8}$.
C_6 has the same sign as b_{41}. If

$$V(x,y)=x^2+y^2+F_3+F_4+F_5+F_6,$$
$$\left.\frac{dV}{dt}\right|_{(8)}=C_6r^6+O\left(r^6\right).$$

When $b_{21}=0, b_{41}>0,$ *the point* $A(0,0)$ is the second-order unstable weak focus.
When $b_{21}=0, b_{41}<0,$ *the point* $A(0,0)$ is the second-order stable weak focus.
Third case: $b_{21}=0, b_{41}=0$.
Given that $P(-x,y)=P(x,y)$ and $Q(-x,y)=-Q(x,y)$, the vector field $(P(x,y),Q(x,y))$ is symmetrical with respect to the y-axis. Hence, point $A(0,0)$ is the center.

6.2 Nonexistence of limit cycle

Theorem 2 If one of the following conditions is satisfied, then closed orbits are absent in the whole plane of system (8).

1. $b_{12}<0, b_{21}>0, b_{41}\geq 0$;

2. $b_{12}<0, b_{21}\geq 0, b_{41}>0$;

3.$b_{12} \leq 0, b_{21} > 0, b_{41} > 0;$

4.$b_{12} > 1, b_{21} \leq 0, b_{41} < 0;$

5.$b_{12} > 1, b_{21} < 0, b_{41} \leq 0;$

6.$b_{12} \geq 1, b_{21} < 0, b_{41} < 0.$

Proof When $b_{12} \neq 0, b_{50} = 0$, let

$$L(y) = b_{12}y - 1$$
$$\left.\frac{dL(y)}{dt}\right|_{(8)} = b_{12}\dot{y}$$
$$= (1 - b_{12})x + b_{21}x^2 - 2b_{12}x^3 + b_{41}x^4$$

Thus, $y = \frac{1}{b_{12}}$ is the line and not the tangent of the trajectory of the system (8). The Dulac's function is $B = \frac{1}{b_{12}y-1}$ (when $b_{12} \neq 0$). Then,

$$div(BP, BQ)|_{(8)} = -\frac{(1 - b_{12})x + b_{21}x^2 - 2b_{12}x^3 + b_{41}x^4}{(b_{12}y - 1)^2}$$

We obtain $div(BP, BQ)|_{(8)} \leq 0$ when the condition (1), (2), or (3) holds.
We obtain $div(BP, BQ)|_{(8)} \geq 0$ when the condition (4), (5), or (6) holds.
When any one of the conditions in Theorem 2 is satisfied, then $sgn\ div(BP, BQ)|_{(8)}$ is definite. Moreover, $div(BP, BQ)|_{(8)} = 0,$ if and only if, $x = 0$. In other words, it is not identically zero in any subregion in the (x, y) plane. Thus, the system (8) has no closed orbits in the whole plane under any one of the cases in this theorem.
Theorem 3 When $b_{21} = 0, b_{41} = 0,$ then the system (8) has no limit cycle.
Proof Since $\boldsymbol{div(PB, QB) \equiv 0}$, so the system (8) has a continuous differentiable integral factor $B(x, y)$, there is no limit cycle. The simulation is presented in **Figure 1**.

6.3 Existence of limit cycles

Theorem 4 The system (8) contains at least one limit cycle around A(0,0) when one of the following conditions is true:

1. $b_{21} > 0,$

2.$b_{21} < 0,$

3. $b_{21} = 0, b_{41} > 0,$

4. $b_{21} = 0, b_{41} < 0.$

Proof The singular point $A(0, 0)$ is an unstable and weak focus point of system (8) under the conditions (1) and (3). Singularity $A(0, 0)$ of the system (8) changes from an unstable weak focus to a stable focus. According to Hopf bifurcation theory, the

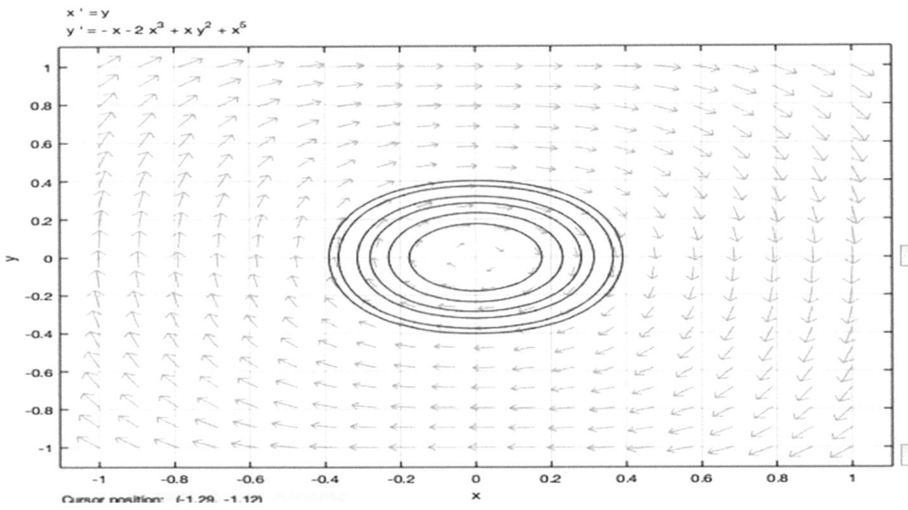

Figure 1.
Center for nonlinear system.

system (8) generates at least one unstable limit cycle around the point $A(0,0)$ under the changes in these parameters.

Under the conditions (2) and (4), singular point $A(0,0)$ is the stable and weak focus point of the system (8). The singularity $A(0,0)$ of the system (8) changes from stable, weak focus to an unstable focus. According to Hopf bifurcation theory, under the changes in these parameters, the system (8) generates at least one stable limit cycle around $A(0,0)$ (point)

Theorem 5 System (8) contains at least two small-amplitude limit cycles bifurcating from the origin $A(0,0)$.

Proof If Theorem 4 holds, then there are perturbations of system (8) yielding two small-amplitude limit cycles bifurcating from the origin. According to Theorem 1, we determine that system (8) has two weak focus points. We denote the number of weak focus as k, and $k=2$. The system (8) is a class quintic; thus, we denote n as $n=5$. We determine that the system (8) has two limit cycles around the $A(0,0)$.

Theorem 6 System (8) contains a maximum of two limit cycles when $b_{12}=0$.

Proof When $b_{12}=0$, the system (8) transforms to the following form:

$$\frac{dx}{dt}=y,$$
$$\frac{dy}{dt}=-x-2x^3+b_{21}x^2y+b_{50}x^5+b_{41}x^4y.$$

This system is equivalent to the following system:

$$\frac{dx}{dt}=y,$$
$$\frac{dy}{dt}=-g(x)-f(x)y,$$

where $g(x) = -x - 2x^3 + b_{50}x^5$ and $f(x) = b_{21}x^2 + b_{41}x^4$. $g(x)$ is odd, and the degree of $f(x) = 4$ is even based on the definition of the local result of the Liénard system and the results of Blows and Lloyd. The maximum number of small-amplitude limit cycles bifurcated from the origin is two.

6.4 Numerical solution

Example 1 If we set the following parameters in system (8): $b_{21} = 1, b_{12} = 1, b_{41} = 1$, and $b_{50} = 1$, then a limit cycle exists around $A(0, 0)$. The simulation is presented in **Figure 2**.

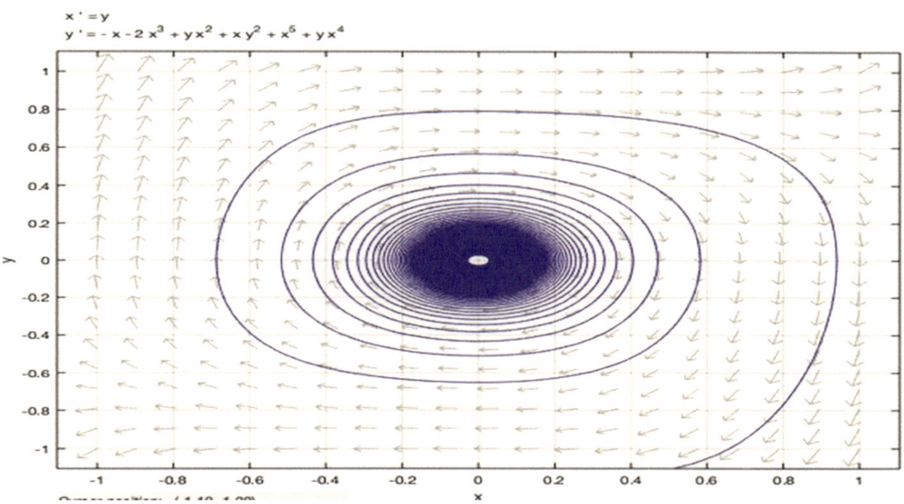

Figure 2.
Existence of the limit cycle for system (8), when $b_{21} = 1, b_{12} = 1, b_{41} = 1,$ *and* $b_{50} = 1$.

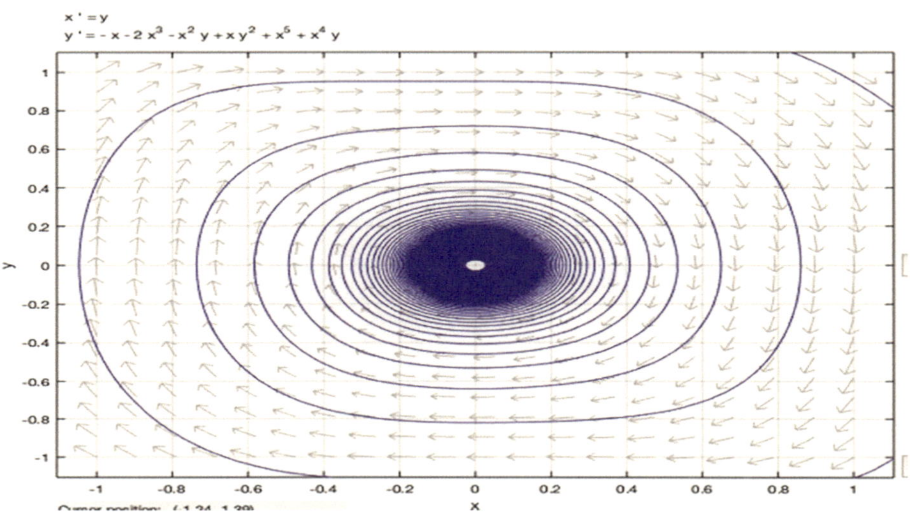

Figure 3.
Existence of the limit cycle for system (8), when $b_{12} = 1, b_{21} = -1, b_{41} = 1,$ *and* $b_{50} = 1$.

Example 2 If we set the following parameters in the system (8): $b_{12} = 1, b_{21} = -1, b_{41} = 1,$ and $b_{50} = 1$, then a limit cycle exists around $A(0, 0)$ according to Theorem 4 of the existence of limit cycle (Case 2). There exists a limit cycle around. The simulation is presented in **Figure 3**.

If we set the following parameters in the system (8): $b_{12} = 1, b_{21} = 0.5, b_{41} = 1,$ and $b_{50} = 1$, then a limit cycle exists around $A(0, 0)$ according to Theorem 4 of the existence of limit cycle (Case 1). The simulation is presented in **Figure 4**.

If we set the following parameters in the system (8): $b_{12} = 1, b_{21} = 0, b_{41} = 1,$ and $b_{50} = 1$, then a limit cycle exists around $A(0, 0)$ according to Theorem 4 of the existence of limit cycle (Case 3). The simulation is presented in **Figure 5**.

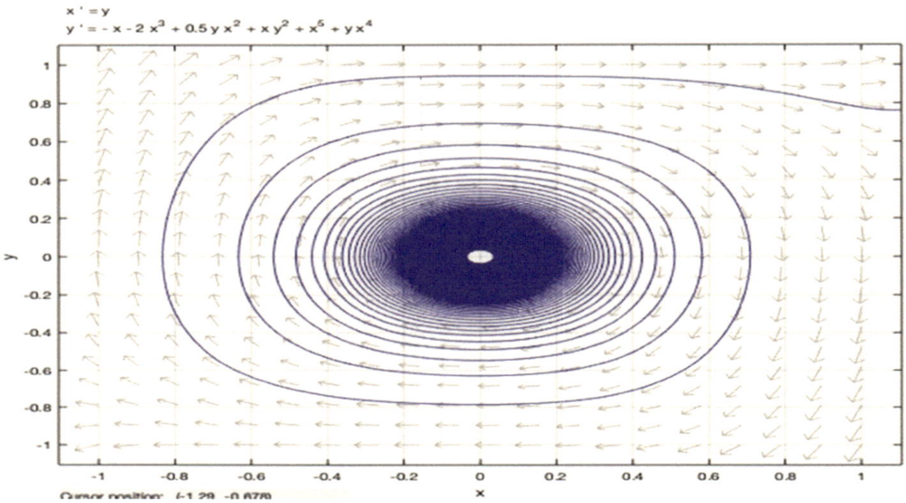

Figure 4.
Existence of the limit cycle for system (8), when $b_{12} = 1, b_{21} = 0.5, b_{41} = 1,$ *and* $b_{50} = 1$.

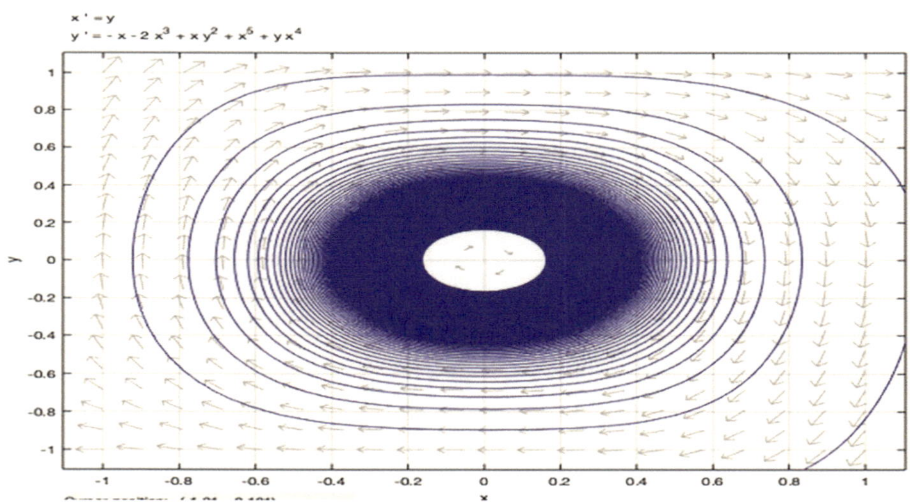

Figure 5.
Existence of the limit cycle for system (8), when $b_{12} = 1, b_{21} = 0, b_{41} = 1,$ *and* $b_{50} = 1$.

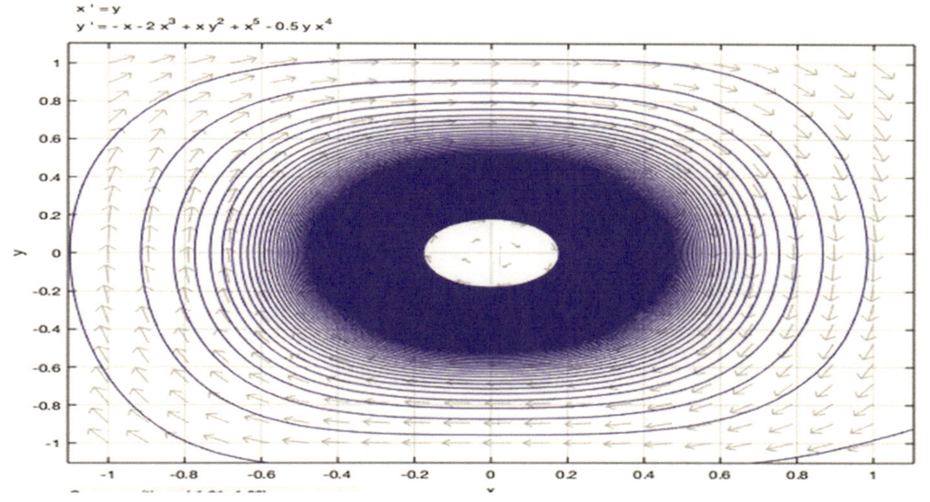

Figure 6.
Existence of the limit cycle for system (8), when $b_{12} = 1, b_{21} = 0, b_{41} = -0.5,$ *and* $b_{50} = 1$.

If we set the following parameters in the system (8): $b_{12} = 1, b_{21} = 0, b_{41} = -0.5$, and $b_{50} = 1$, then a limit cycle exists around $A(0, 0)$ according to Theorem 4 of the existence of limit cycle (Case 4). The simulation is presented in **Figure 6**.

7. Conclusions

We investigated the existence of the limit cycle and used the formal series method to determine the center-focus in this chapter.

We establish the sufficient conditions for the existence of the limit cycles in system (8) that bifurcate from the equilibrium point using Hopf bifurcation theory and discuss the nonexistence of closed orbits using the Dulac function. Some examples were provided for illustration.

A number of interesting problems arise from this work. Future research in this area could include finding a general solution formula that can be used for algebraic equations of higher degrees and the calculation of system singularity. Such problems offer researchers a wealth of opportunities to expand this new area.

Acknowledgements

The authors are grateful to both reviewers for their helpful suggestions and comments.

Conflict of interest

The authors declare no conflict of interest.

Thanks

The authors are grateful to Prof. Ivana Barac for her help and cooperation.

Author details

Sarah Abdullah Qadha[1,2*], Muneera Abdullah Qadha[1,2] and Haibo Chen[1]

1 School of Mathematics and Statistics, Central South University, Changsha, China

2 Faculty of Education at Al-Mahweet, Department of Mathematics, Sana'a University, Al-Mahweet, Yemen

*Address all correspondence to: sarah-qadha@csu.edu.cn

References

[1] Poincaré H. Mémoire sur les courbes définies par une équation différentielle (II). Journal of Maths and Pures Applications. 1882;**8**:251-296. Available from: http://eudml.org/doc/235914

[2] Poincaré H. Mémoire sur les courbes définies par une équation différentielle (I). Journal of Maths and Pures Applications. 1881;**7**(I):375-422

[3] Poincaré H. Sur les courbes définies par les équations différentielles (III). Journal of Maths and Pures Applications. 1885;**1**(4):167-244

[4] Poincaré H. Sur les courbes definies par les equations differentielles(IV). Euvres de Henri Poincaré. 1985;**1**(Iv):85-162

[5] Ye Y-Q et al. Theory of limit cycles. The American Mathematical Society. 1986;**66**

[6] Han M, Lin Y, Yu P. A study on the existence of limit cycles of a planar system with third-degree polynomials. International Journal of Bifurcation and Chaos. 2004;**14**(1):41-60. DOI: 10.1142/S0218127404009247

[7] Mattuck A. LC. Limit Cycles. Journal of Differential Equations. 2011;**2011**:1-6

[8] Cao J. Limit cycles of polynomial differential systems with homogeneous nonlinearities of degree 4 via the averaging method. Journal of Computational and Applied Mathematics. 2008;**220**(1–2):624-631. DOI: 10.1016/j.cam.2007.09.007

[9] Cioni M, Villari G. An extension of Dragilev's theorem for the existence of periodic solutions of the Liénard equation. Nonlinear Analysis: Theory, Methods & Applications. 2015;**127**:55-70. DOI: 10.1016/j.na.2015.06.026

[10] Benterki R, Llibre J. Centers and limit cycles of polynomial differential systems of degree 4 via averaging theory. Journal of Computational and Applied Mathematics. 2017;**313**:273-283. DOI: 10.1016/j.cam.2016.08.047

[11] Giné J, Llibre J, Valls C. Centers for the Kukles homogeneous systems with odd degree. The Bulletin of the London Mathematical Society. 2015;**47**(2): 315-324. DOI: 10.1112/blms/bdv005

[12] Giné J, Llibre J, Valls C. Centers for the kukles homogeneous systems with even degree. Journal of Applied Analytical Computers. 2017;**7**(4): 1534-1548. DOI: 10.11948/2017093

[13] Bautin N. On the number of limit cycles appearing with variation of the coefficients from an equilibrium state of the type of a focus or a center. Maths Science Net. 1952;**1**(1939):2414

[14] Lloyd NG, Pearson JM. Computing centre conditions for certain cubic systems. Journal of Computational and Applied Mathematics. 1992;**40**(3): 323-336. DOI: 10.1016/0377-0427(92)90188-4

[15] Hong Z, et al. Limit cycles for the Kukles system. Journal of Dynamical and Control Systems. 2008;**14**(2):283-298. DOI 10.1007/s10883-008-9036-x

[16] Christopher CJ, Lloyd NG. On the paper of jin and wang concerning the conditions for a Centre in certain cubic systems. The Bulletin of the London Mathematical Society. 1990;**22**(1):5-12. DOI: 10.1112/blms/22.1.5

[17] Chow SN, Li C, Wang D. Normal Forms and Bifurcation of Planar Vector Fields. Cambridge University Press. 1994

[18] Munoz R. Introduction to Bifurcations and the Hopf Bifurcation Theorem for Planar Systems. Colarado State University; 2011. pp. 11-14

[19] Lynch S. Dynamical Systems with Applications Using MapleTM. Springer Science & Business Media; 2009

[20] Zhang Zhi-fen HW. Qualitative Theory of Differential Equations. United States of America: American Mathematical Society; 1991